EDWARD HANCOX is obse:
He has written about Iceland fo
sites, including *Iceland Review* :
In 2013, he released his first boo... *Iceland, Defrosted* to a sell out crowd at the Embassy of Iceland in London.

His obsessive streak has led to an fascination with the Atlantic puffin and particularly their plight for survival. The love affair endures to this day.

Edward resides in Shropshire, England and is a proud Salopian. He spends his spare time writing, enjoying the outdoors and drinking real ale. Sometimes all at once.

EDWARD HANCOX

EVERY LAST PUFFIN

SilverWood

Published in 2021 by SilverWood Books

SilverWood Books Ltd
14 Small Street, Bristol, BS1 1DE, United Kingdom
www.silverwoodbooks.co.uk

ISBN 978-1-80042-031-1 (paperback)
ISBN 978-1-80042-032-8 (ebook)

British Library Cataloguing in Publication Data
A CIP catalogue record for this book is
available from the British Library

Page design and typesetting by SilverWood Books

This book is dedicated to
Nichola, Lily and Ella for all the love and patience,
and to my close friend SJ for believing in me
and feeding me cream teas.

Special thanks to Mum, Dad, Lisa, Ally, Bec Herdson, Blair, Cymbaline Wilhite, Daniel Scheving, Einar, Jack and Oliver Hart, Katrina Gilman, Kevin, Liz, Martin, Michelle Clarke, Paul, the Singh family, Tracy and Vessela.

Contents

Every Last Puffin

- Unst
SHETLAND
- Sumburgh Head

- Westray
ORKNEY

NORTH
ATLANTIC
OCEAN

- Dunnet Head

- Handa

- St Kilda
- Shiant Isles

NORTH SEA

SCOTLAND
- Fowlsheugh

- Treshnish Isles

- Isle of May

- Farne Isles

EDINBURGH

- Rathlin Island

- Coquet Island

NORTHERN
IRELAND

- Bempton

BELFAST

IRISH SEA

DUBLIN

- Anglesey/Ynys Môn

- Bardsey/Ynys Enlli

IRELAND
WALES
ENGLAND

- Skomer Island
CARDIFF
LONDON

- Lundy

CELTIC SEA

N

ENGLISH CHANNEL

- Annet
ISLES OF SCILLY

0 _____ 100 miles
0 _____ 100 km

INTRODUCTION

An Improbability of Puffins

Puffins are often affectionately called the penguins of the north, clowns of the sea or sea parrots. A group of puffins can be a burrow, an improbability or even a circus. With their rainbow-coloured beaks, bright orange feet, stout black-and-white bodies and mournful faces, puffins are the perfect rock stars.

There's bad news though. This enigmatic little seabird is on the 2020 International Union for Conservation of Nature (IUCN) list of species 'vulnerable to extinction'. Although there are around 580,000 puffins in the United Kingdom, they are on the red list globally due to their continued decline. Some figures from the Royal Society for the Protection of Birds (RSPB) estimate further declines in the UK puffin population of between 50 and 70 per cent by 2065. The outlook is not good.

The IUCN details a number of reasons why the puffin is struggling. Rising sea temperatures attributed to climate change have altered the distribution and availability of key puffin food sources, such as the sand eel. Puffins are having to travel further and further to find food, and sometimes don't find it at all. Sand eels themselves are subject to overfishing, with teams of trawlers removing vast quantities from our seas. An increase in extreme weather events linked to climate change has proven deadly for puffins out at sea over the winter months. Rats and other predators, often inadvertently introduced to remote islands by humans, have invaded puffin colonies, predating precious chicks and eggs. Human activity has also threatened the puffin in other ways. Oil spills and plastic pollution are having a detrimental impact on our seabird populations, and puffins are still being hunted for food in countries such as Iceland and the Faroe Islands. It is a bleak picture for the little puffin.

Weighing little more than a tin of baked beans, the Atlantic puffin is the typical puffin, the one that adorns books, cartoons and, somewhat bizarrely, a breakfast cereal in the United States. It's the one we have in the British Isles, and it also breeds on the coasts of Iceland, Norway, Greenland, Canada, Russia and the Faroe Islands. It's the one that we're concerned with throughout this book.

This story starts in Iceland, specifically in Heimaey on the Westman Islands. There I met a local celebrity, a rescued puffin called Tóti, who was showing off his colourful bill and orange feet as he padded around the Sæheimar Aquarium. I much prefer seeing puffins in their natural environment, but I was there to learn about puffin rescue and Tóti was a perfect example of that. Named after a local footballer, he was one of many puffins rescued by Heimaey's children every season. When the pufflings – juvenile puffins, and the cutest term in the world – emerge from their burrows for the first time, they get confused by the town's lights and head towards them rather than out to sea. So, local children form puffin patrols and gather up the stragglers each night, placing them in cardboard boxes before releasing them safely into the sea. Tóti was one of these rescued pufflings, but he was weak and it was so late in the season that he couldn't be released. Instead, he was being looked after in the Heimaey aquarium. He led a carefree life, able to wander – and poop – where he liked, giving much pleasure to the centre's staff and visitors.

Outside, I managed to catch up with a puffin patrol – two Icelandic girls of about twelve who were pulling a couple of cardboard boxes on a cart on wheels. It was not that late, but the girls were taking their duties seriously during the long summer holidays. Heimaey is rightly proud of its puffins; even the street signs and benches have puffin faces, the shops are full of puffin souvenirs and the Chinese restaurant has a mural of a puffin with chopsticks. I asked the girls if I could look inside the boxes at the pufflings. 'Jæja,' came the bemused reply, as if they couldn't quite understand why a strange English chap wanted to look at pufflings.

The first box was opened to reveal a worried-looking puffling of around two weeks old. Thinner than I'd expected, it had a white breast, grey plumage, a sharp, black

15

beak and intelligent eyes. The box contained luminous yellow excrement that caused the girls to wrinkle their noses. They told me they'd found the puffling a couple of streets away, behind a car tyre. The second box contained a much younger bird, probably only a few days old. This was not much more than a sphere of downy grey fluff with two eyes and a short beak. As cute as it was, I was worried about its chances of survival and I was glad the girls had found it, especially when they told me they'd seen it being eyed up by a local cat. 'We scared it off though,' they said proudly in that Icelandic/American accent learnt partly from Nickelodeon. I politely declined to hold the pufflings, but we agreed that I'd accompany the girls to the aquarium to weigh and record the size of the birds.

Iceland and, particularly, the Westman Islands are seen as a global stronghold for the puffin, and yet, since 2003, the islands have seen a massive decline in the number of birds. Some Westman Island cliffs, once covered in puffins, have remained sadly empty. Even at Stórhöfði, a large colony in the south of Heimaey where you can closely watch puffins from the shelter of a green wooden hut, I had seen a noticeable difference in the number of puffins. The Westman Islands have a long tradition of puffin hunting, with hunters using a large nets on poles to literally 'fish' the puffins from the sky. Puffins were a staple food on the islands. Nowadays, hunting is limited to two or three days a year to protect the puffins, and even then, the hunters are less than successful. Hunters themselves recognise the trouble that the puffin is in. It would be hard not to see, from the barren and desolate cliff tops to the pufflings found dead in their burrows from starvation. Climate change is having a devastating impact on the Icelandic puffin population.

Later that day, I returned to the aquarium to release the birds. Or bird, in fact. The youngest puffling was as fragile as I'd feared and needed to spend some time recuperating and being cared for before it was fit for release. Out on the black sand of the volcanic beach, the box was opened again. The girls let me reach in and take hold of the petrified bird with both hands. I felt its heart racing, and mine, so I didn't linger. Removing the puffling from the box, I gently tossed it towards the sea. It responded well and was clearly pleased to be free of me. There was a brief fluttering of new wings and then a soft landing on the sea. The puffling gathered itself before ducking underwater. It was gone. The girls were already halfway up the beach on the search for more lost pufflings. Me, I was smiling away, proud to be the newest member of the puffin patrol.

Atlantic puffins are in trouble, but I can't save them one bird at a time, as I did in Iceland. That's what got me thinking. What if I was to highlight the plight of puffins in the UK? Ask your friends and neighbours: I bet very few have seen a puffin at all, let alone in the UK. Puffins breed in the UK, but colonies are under pressure and some have disappeared altogether. What if I was to try and visit each colony in the UK?

And so, a loose plan was born – I would visit puffin colonies across the UK, seeing as many puffins as I could. Fortunately, they happen to live in some really spectacular locations. I would try to learn more about this bird before it was too late and, at the same time, I'd learn about the flora and fauna with which they share the clifftops on their brief trips to land.

Here, then, are the stories of what I found, including the part where I took a dunking in the cold Atlantic and an account of how I tried not to lose my stomach, or teeth, on the roughest of seas. The book opens in early April, when puffins make their annual landfall in the UK, and concludes in late July, when they start to leave our shores. In between, as I meander around the UK between their various colonies over the course of several years, I discover what puffins get up to, why they do what they do, the struggles they are facing and, in the words of a very famous children's book, why, exactly, there is nothing like a puffin.

1

It's the Little Things

April

Fowlsheugh, Kincardineshire, Scotland

The Caledonian Sleeper had been in the news. It was about to get brand-new trains to shuttle passengers from London Euston to the Scottish Highlands. They were calling it a hotel on wheels, and it would have 347 toilets, en-suite cabins and a Jacuzzi. I made the Jacuzzi up; that would never work. There were artist's impressions of trains passing through sweeping glens, chugging over majestic viaducts and stopping conveniently at golf courses and shooting estates. Every member of staff was wearing tartan and a welcoming smile, and passengers were served rare Scottish venison, Cullen skink and a variety of the finest whiskies. After dinner, travellers could either retire to plush suites with a fold-down service and pillows graced with deluxe

Scottish chocolates, or stay up to watch the Northern Lights. Everyone would wake next morning to find a member of staff serving steaming-hot coffee and an Ayrshire bacon roll.

The reality was nothing like that. I'd wanted to do this trip for years, so I was excited to board the 2350 from Crewe to Stonehaven. The station was deserted except for confused herring gulls wheeling between the car park lights, and furious freight trains rattling by. Vending machines glowed in the darkness and long-forgotten bicycles clanked in their racks as the wind blew through. I got on, and simultaneously stepped back into 1985. Nothing had changed. The corridors were narrow and rickety, painted British Rail grey and scuffed throughout. I found my luxury reclining seat. It was threadbare. The material had separated from the seat and hung in dirty fronds. It didn't recline, a serious omission for a reclining seat, so I jammed my knees into the seat in front. I felt my kneecaps dislodge but figured the pain would probably stop somewhere near Carlisle and I could seek medical attention at Dundee. At least it discouraged me from visiting the toilets. These sent the occasional whiff into my carriage – one was full of faeces, the other had a sink overflowing with dirty water. I tried to ease my way up the train to first class – surely, they'd have better toilets – but I was led, or should I say escorted, back to my carriage of grunting fellow adults enduring seven hours of sleep deprivation by a gruff member of staff who I didn't recall from the advertising. I was not even given any shortbread.

The train uncoupled at Edinburgh, splitting in three for Inverness, Aberdeen and Fort William. It was 3am, and there was no sign of the turn-down service or any sleep. Once the uncoupling had finished – no allowances were made for those in the Land of Nod – we sat, for reasons unknown, at Edinburgh, watching dozy pigeons and the arrival of the early-morning papers. Sleep clearly wasn't going to happen, so once we were out of the suburbs, I watched the sun rise over a frost-sprinkled Scotland. I saw a couple of red deer, which I was thrilled about, but there was no one to share this with. Spring hares bounded across empty fields. Brilliantly, staff had stopped waking passengers at stations, so as we hurtled further north there were exclamations of annoyance as people woke to find they were long past their stop. It was a great system. I was glad to alight at Stonehaven. I hope the new carriages and smiley, tartan-clad staff arrive soon on the Caledonian Sleeper and that it regains the glamour that has long since faded.

Stonehaven seemed nice enough, and I found a place to serve me a full Scottish breakfast, which I devoured to ward off the tiredness. There is something extra Scottish about potato scones and square sausages that always puts me in a good humour, even if this doesn't do much for my arteries. Plus, there was a beautiful puffin mosaic on the floor outside the Ship Inn, which I took as a good sign. The puffin was coming in to land, feet and wings outstretched and backlit by the sun. It was way too good to be trampled on and covered in regurgitated fish suppers, in my view. I took a sneaky photo of it to admire later on.

The mosaic puffin was an Atlantic puffin, like the ones I'd come all the way up here to meet. But before we get to the Fowlsheugh RSPB reserve, I'd like to introduce you

21

to some of its far-flung relatives, beginning with the horned puffin. Horned puffins live around the northern Pacific coasts of Alaska, Canada and Siberia. They are black and white like their Atlantic cousins, although with rounder, light lemon-yellow beaks, with a deep orange tip. They are also slightly larger in size. They have a prominent black dash – or horn – above each eye, which makes them look a bit startled, as if you've crept up on them unannounced. Unlike the Atlantic puffin, horned puffins are not currently under threat.

The tufted or crested puffin shares the Pacific coasts with its smaller horned cousins. Nearly all black, the large tufted puffin has a distinctive yellow or golden tuft protruding from its head, a thick orange beak and a white facemask that makes it look a bit, well, sinister. I can't help thinking of Feathers McGraw, the evil penguin in *Wallace and Gromit*, when I see one. Somewhat bizarrely, a tufted puffin was spotted on the Swale estuary in Kent in September 2009. This first known sighting in the UK caused a big 'twitch' and prompted much speculation as to why the tufted puffin was thousands of miles from home. With the exception of that Kent bird, tufted puffins are not in trouble.

Although arguably not a puffin at all, the rhinoceros auklet is a distant relative. It doesn't look at all like the other three puffins, being dark and drab, with the eponymous protrusion above its small orange beak. I've also heard it called a unicorn puffin, which sounds like some sort of mythical creature, but that is a stretch too far for such a sombre-looking bird. It has a range across the north Pacific, and is not threatened in any way. The rhinoceros auklet is, by some distance, my least favourite of the puffin family, in part for its sharp, evil-looking beak and beady, angry eyes. It appears to be frowning in disdain the whole time; maybe it's angry that it's not a cute puffin. A unicorn puffin it is certainly not.

The Atlantic puffin is hands down my favourite, and I was keen to get on with my trip to RSPB Fowlsheugh. The reserve is a few miles south of Stonehaven, near a clutch of houses and a natural harbour called Crawton. I decided to catch the bus for this relatively short journey, and I was ridiculously excited that it was a double-decker. I went to the top, of course, and sat right at the front so I could pretend I was driving. Sometimes it's the little things. I was dropped off at Uras Cottages, which sat either side of the road, ominously empty and in various states of disrepair. The plastic sheets patching up the smashed windows were being pulled at by the gusting wind, and a loose drainpipe moaned in reply.

I made my way down the single-track lane towards Crawton. It still felt wintry here; the trees were not yet in bud and the thick moss covering the stone walls was fat with moisture. Wind whipped across barren fields that crows and an out-of-place oystercatcher were picking over for worms. The only signs of spring were the bright blooms of wild daffodils and a valley of yellow gorse. Both were welcome sights against the grey sky and my tired, slightly befuddled mind. A cock pheasant broke from the hedgerow in front of me, noisily faking his surprise, and I was instantly transported back to being a twelve-year-old lad in rural

Shropshire. Before I was old enough to know better, I used to go beating for a local landowner, flushing out pheasants from woodland to be shot, in return for a crisp £10 note and an illicit can of beer. At lunchtimes, we'd sit around an open fire, clothes steaming dry, as tales were told and plans were made for the wood with the richest pickings. The gamekeepers held court as an old-timer who knew my grandad looked out for me, all the while eating thick cheese sandwiches and biting chunks out of an onion like an apple. Sometimes I'd see woodcock while we were out in the woods and I'd keep deliberately quiet. Those poor creatures were highly prized – pan-fried with their beaks pushed into their own bodies and placed on a piece of toast. I'd let them fly past me, away from the row of guns.

I carried on to Fowlsheugh down twisty lanes with hedgerows emerging from winter dormancy. Rabbits scuttled away from me, surprised at my presence. The RSPB reserve runs for a mile or so along the coast with cliffs some eighty metres high. It was marked only by a couple of signs and a wooden stile; no fancy visitor centre here. A stream wound through a small valley on its way to the sea, blown from the clifftops by the fierce wind in a rough approximation of a waterfall. Lazy herring gulls paddled in the stream and chuckled at me as I passed. Common scurvy grass provided a luxurious carpet of early white flowers.

I approached from the southern end and, as I made my way north, I passed plenty of herring gulls, fulmars and kittiwakes; guillemots and razorbills were present too, but not up to their full numbers. It was still early in the season and the sea below the cliffs was full of flocks of birds, undecided whether it was yet time to come on land. It was causing much chatter among the razorbills in particular. Around 130,000 seabirds choose to nest at Fowlsheugh, and I could see why. This corner of Scotland is completely undisturbed. From the cliffs where I stood looking out to sea and north to Dunnottar Castle, which appeared like a film set from a cancelled Netflix series, I couldn't spy another human being.

The sea was blue and calm beneath me and the sun had started to show through. A couple of ships were heading south towards Dundee or Edinburgh. I reached Henry's Scorth, a fierce chasm, or geo, cut into the rock, each face jagged and angry. It was a clear favourite with the birds though, with cliff ledges filling quickly and the best real estate already being claimed for the season. The cries of guillemots and razorbills echoed off the steep walls.

Checking my map, I noticed that I was looking at Hope Cove and that the next dog-leg in the path was called Fail-You-Never. This had to be a sign that the puffins were not far away, surely? As if in agreement, I heard the unmistakable noise of eiders below me, creaking and groaning, and then I found an actual sign, a single post in the ground onto which someone had carved and painted a tiny puffin character. The wooden post had seen better days, with the sea air clearly having taken its toll and lichen claiming the top as its own.

I'd been told that the puffins were to be found at the north end of the reserve, near the newly built hide, so I continued on, accompanied by rock pipits and wheatears who watched my every move. A yellowhammer hopped from post to post, singing from each new stage like a *Britain's Got Talent* contestant.

The new hide was a real beauty; made from blocks of local stone, with a big thick door to keep the draught out and an expansive wooden seat that looked out across the sea. A plaque stated that the hide had been built 'in memory of George W Anderson and his wife Dr Moyra SM Anderson, whose generous bequest provided for the erection and maintenance of this shelter to ensure that others could enjoy, in comfort, the spectacular beauty of an area which gave them both so much pleasure'. I couldn't agree more. This was an area of spectacular beauty, and it was the only place around that provided any shelter. This coast is one of the windiest in the UK, so I'm sure the hide will be put to good use and many a person will silently thank the Andersons for their gesture. It is also no coincidence that the hide was built overlooking Dovethirl Cove, which is where twenty or so puffins breed on the cliffs below. It's a glorious spot.

I wasn't the only one on the path though. Not like I'd thought. Up ahead of me was a stoat. It wasn't long before it spotted me, but I'd got a head start and managed a good look. The old saying, 'How do you tell the difference between a weasel and a stoat? A weasel is weasel-y recognisable where a stoat is stoat-ly different,' went through my mind. Actually, it's not so difficult. The real difference is the stoat's tail, which, just like the one in front of me, has an obvious black tip. The stoat took a good look at me to satisfy its own infinitely curious nature, before bouncing off into the undergrowth.

Signboards dotted along the clifftops informed visitors of what was going on around them and on the cliffs and sea below. The one about fulmars was entitled 'Stinky Spitters' and I was amused to see that a bird had performed a dirty protest on it. I only hoped it was a fulmar. Other signboards communicated less good news – one stated that the last count of guillemots had revealed a decline in their numbers and made the link back to changes in food supply and the climate.

A little further down the path and over a small bridge and I was in puffin territory.

The puffins were in stark contrast to their surroundings. Spring had yet to reach this cove, and the slopes preferred by the puffins remained barren save for flowerless patches of thrift and brown tufts of grass that had failed to survive the cold winter. There was the odd splash of white from early-nesting kittiwakes, but you couldn't miss the fireworks display produced by the distinctive colours of the puffins. Although perfect camouflage in the sea, the black and white stood out a mile here, especially the fulgent white of their chests and the freshly squeezed orange of their legs. The morning sun caught the reds and blues of their beaks, setting them on fire too.

Only two or three puffins had come on to land, making a head start on finding their burrows. Though they were few, it made me genuinely happy to see these first puffins of spring in this remote corner of Scotland. They had chosen a spot with a shallow cave halfway down the cliff face and with a length of discarded rope nearby that seemed to act as a handy plaything.

A few razorbills shared the same place but looked under-stated alongside the puffins, wearing dinner jackets rather than extravagant ball gowns. A couple of them began brazenly copulating in front of the puffins and me. None of us were interested in this vulgar display. It was over in seconds anyway. Puffins rarely copulate on land, it should be noted, but save such antics for the sea prior to arriving back in the colony.

Bridled guillemots poked their heads over the top of the cliff to observe me like bespectacled teachers. It looked for all the world as if someone had drawn on their spectacles, using the finest brushes and white gloss.

There were a couple of puffins further around, sharing a precarious ledge of rock. A freshly reunited pair, they did some gentle billing (rubbing of beaks) and head jerking, and then set about preening themselves, a welcome chance to sort out feathers battered and bedraggled after a winter at sea. Like other seabirds, puffins have a gland at the base of their tail, which secretes an oil to keep their feathers waterproof. They need to preen frequently to maintain their feathers in good condition.

My eye was drawn to the sea, where three little dots bobbed not far from where I'd seen a couple of eiders earlier. Could they be little auks? I'd only previously seen little auks in Iceland, in the Westman Islands, where they lined the cliffs like miniature penguins. I delved into my bag to retrieve my binoculars and then tried to find the auks again on the gently surging sea. Yes, got them. They were little auks, and I was thrilled.

This little nature reserve was certainly full of hidden treasure. Little auks, or dovekies as they are sometimes known, are just as it says on the tin. They are small – the size of a blackbird but chubbier – and a member of the auk family. They have similar, but smaller, rear halves to their puffin and guillemot cousins – black and white with a slightly tufted tail – with soot-black heads and shiny button eyes. They are fairly frequent visitors to Scotland; although whether they want to be or not is another thing entirely, as they are often blown off course by strong winds and storms. These three looked happy enough though, rocking back and forth on the gentle swell. They'd probably seen much worse weather during their travels.

I returned to Fowlsheugh the following day, hoping to catch up with the puffins and do some sea-watching from the shelter. It's a good spot for bottlenose and common dolphins and, if you're lucky, minke whales and harbour porpoises. It was overcast and much colder than the previous day and, as soon as I got onto the clifftops, I noticed the difference. The birds were quieter and there were fewer of them on the cliff face, save for the kittiwakes, which were huddling together. A herring gull announced my presence with a raucous shout-out. I made my way to Dovethirl. Today, I was not alone. There was a chap standing near the shelter, just off the path. As I approached, he turned to me and waved with one hand, clutching a camera with the other. He was an older man, smartly dressed, and with a red face. 'There's a puffin!' he said excitedly. 'I've never seen one before, after all these years!'

There was a solitary puffin sitting on a tuft of grass some thirty metres from us. The man was Scottish but had lived in England for most of his life. In his later years, he'd decided that he needed to come back to Scotland. One of the things he'd always wanted to do was see a wee Scottish puffin. I chose not to mention that the previous day there'd been about twenty of them in this very place, or that there were still places in the UK where you could get so close to puffins that they would play with your shoelaces. Instead, we sat and shared the moment. The joy on that old man's face at having seen a puffin was something to behold. A smile had spread across his cheeks and refused to leave. Puffins bring people joy. I'd go one step further and say that puffins are good for people's mental health – mine as well as that of my newfound companion.

Fearing that the Caledonian Sleeper would not have improved any in the few days since my trip north, and still having nightmares over the toilet situation, I took a ScotRail train back south. A nice lady pushed the trolley and supplied me with super-sweet Orkney fudge and scalding hot tea. When she checked my ticket, she handed it back with a tiny penguin figure punched through it. I noticed, and thanked her. 'Aye,' she said, 'It's the little things that make people smile.' She was right, of course; little things like a penguin or a puffin.

2

Cappuffinccino

April

Bempton, Yorkshire, England

Bempton Cliffs were a breath of fresh air, delivered with a fierce wind that blew directly into your nostrils and mouth whether you wanted it or not. It was good, healthy stuff, fetched straight from the cold sea, the sort recommended by grandparents after Christmas lunch or by eye-dabbing ladies at funerals.

The same wind was being harnessed by several large turbines that I saw dotted across the East Riding skyline, courtesy of a series of bank holiday road closures that had me weaving around Beverley and Driffield like a pollen-drunk bee. Close up, the wind turbines were fearsome things, much bigger than I'd anticipated and resembling the front nose of an aircraft. Their aerospace design and sleek lines appeared at odds with the ancient landscape on which they sat. Even the daffodils at their feet, still in full bloom in April, seemed to be shaking their heads at the intrusion.

The wind buffeted the car, causing me to drive in what must have looked a somewhat peculiar manner. Well, that's my excuse anyway. It was early though, so there was no one else around. I didn't see another soul for miles, and it felt as if I had this corner of Yorkshire all to myself; maybe it was the wind, or the unseasonable cold that had enveloped the whole country in a last kiss of winter. In Iceland, they say that if winter and summer are frozen together, the summer will be especially good. If this was the case, we were due a scorcher.

The RSPB centre had not yet open for the day, but there was access through a side gate that led to a footpath and then on to the cliffs. The footpaths were well maintained by the RSPB – I've seen motorways and A-roads with less impressive tarmac. I ambled along the path to the cliffs, the wind whipping at my clothes as I passed flat grassland with lush green hedges and the odd flash of yellow gorse. The sky was bright and painted with lines of grey and pink on a white canvas. I could hear meadow pipits and linnets in the nearby fields, their spring racket audible over the wind, and a pied wagtail danced up the path in front of me, clearly thinking he was a stand-in Bempton guide.

Bempton Cliffs just kind of happen; there is no rise or incline, it's all mainly flat and then the towering cliffs simply drop viciously into the cold sea. It's quite a surprise. At Grandstand, a steep grass slope quickly gave way to the sheer drop of the cliffs, the white, grey and cream faces of which were streaked with the odd line of orange and black.

The sea was surprisingly calm, with little to no wave activity, and I could clearly see the wind rippling the surface like fluffy potato on the top of a cottage pie.

It was my first time at Bempton and, although I was there to see the puffins, I was taken aback by the vast number of birds. Guillemots and gannets wheeled and arced through the air, while beneath me the sea was speckled black with floating birds trying to catch their breakfast. I had earlier had a McDonald's breakfast with a team of match fishermen heading out to a contest, and I couldn't help but make a comparison. At least the fishermen didn't leap off the tables to get their breakfast, although I couldn't say whether they or the birds were the loudest.

There were a few other photographers present – the serious variety who meant business. Green Gortex and camouflaged long lenses were in abundance, making me feel a little insecure about my tiny notepad and point-and-shoot camera. I found one such gentleman on Mosey Downgate, a viewpoint recently erected by the RSPB in fashionable yet ecologically sound timber. Hunched over his scope, which was securely fastened to the fencing, and with another long lens slung over his chest with webbing, he was deep in concentration, trying to get his best photos before the tourists and casuals – like me – turned up and ruined his peace, or worse, his shot. He probably didn't fully appreciate, then, my strangled declaration of 'Yes!' when I spotted two puffins huddled together on a jutting ledge about ten metres from the viewpoint. He might have scowled at me and started to unscrew his scope, but I didn't care. It was only April, but there were puffins here, and I was so pleased to see them.

I managed to prevent myself from doing the fist-pump favoured by goal-scoring footballers. Puffins. Were. Here.

The two had their perfectly black backs to me and were side by side. I could just see their white bellies tucked up beneath them and, in synchronicity, they angled their heads to the left, out of the wind. I could clearly see their colourful beaks, their cute eyes and the last remaining smudges of winter plumage on their cheeks. Puffins lose the brightly coloured platelets, or scutes, from their beaks in the winter, leaving smaller, black ones to prevent them becoming targets on the sea. For the same reason, they also become grey around the eyes, instead of Dulux white. Puffins also moult during their long winters out at sea, leaving them unable to fly until their feathers fully return.

Patrolling the skies and dotted around the cliffs were fulmars. Fulmars are noisy, unforgiving birds with a loud cackle of a call. They are often found alone and are not sociable creatures. They spit a bright orange stomach oil at anything that is brave enough to get close. It stinks. It's the most putrid, fishiest thing you can think of. I know this from having received a dose while trying to observe an Arctic fox in the Westfjords of Iceland. The fulmar's spitting antics are designed to mat the feathers of other attacking birds but, seeing as I don't have any feathers, I was OK, apart from a ruined shirt and being made to take an emergency bath in a mountain stream before being allowed back into the car. The stomach oil also provides the bird with its name, from the Old Norse *fu'll*, which means foul, and *ma'r*, which means gull. That's right – foul gull.

Just above the foul gull was a ledge and a small outcrop of tumbling grass, holding on to the chalk with its fingertips. I was just thinking how tempting it might look to a puffin, when one flew in and landed directly next to the grass. The puffin stood proud and tall, showing off his washing-powder-white chest to the world. He then began pulling at the grass with his beak, presumably starting the chore of nest building. He seemed to concentrate hard on this task before flying gracefully from the ledge in a beautiful arc, then twisting and twirling to the sea below. I followed him as far as I could, before I lost sight of him as he joined the mass of birds on the water.

My hands were freezing, and other people had started to arrive, so I decided to head back to the centre and, more importantly, a hot coffee. In the centre, puffins were clearly in fashion. There were puffin information boards, a fibreglass puffin statue, welcoming visitors to Bempton, and puffin masks for children to colour in. There was also the chance to observe a live puffin, via CCTV, on several widescreen TVs, like a gentler version of *Crimewatch*. The shop was a sight to behold, full of stuffed puffin toys in a variety of designs, puffin bags and tea towels (I am so tempted to include my 'teat owl' joke here), mugs, coasters, key rings, zip pulls, T-shirts, caps, badges and pens, all with puffins on. There wasn't a single fulmar product in the whole place. I considered raising this with the duty manager but decided instead to warm my hands around a very passable cup of boiling coffee and chow down on a sickly-sweet muffin.

Coffee and Bempton go hand-in-hand. In the winter of 2013, puffins off the coast of Yorkshire had a particularly bad time, with a series of storms preventing the birds at sea from getting to their food and ultimately leading to starvation. Hundreds of deceased puffins were found on local beaches in the space of a few days. This sad occurrence is called a puffin wreck. The RSPB estimated that 10 per cent of the Bempton puffin population may have died and immediately set about raising the alarm. This is where the coffee comes in. To promote the puffins of Bempton and entice people to come and see them, and donate handsomely to the RSPB, a local coffee shop owner – Richard Burton from Buckton – came up with the idea of sprinkling chocolate onto cappuccinos through puffin-shaped stencils. He called it a cappuffinccino. Richard, we salute you.

The RSPB has pioneered other means of helping puffins, including the award-winning Project Puffin. Since 2017, the RSPB has asked members of the public to send in photographs of puffins, particularly puffins carrying fish, so that it can monitor which fish are being fed to pufflings and to better understand where issues may exist. The public has taken this to heart, with thousands of photos submitted by 'the puffarazzi' since the project started.

Disappointed to see that the cappuffinccino was not on the menu, I finished my cup of regular coffee and wandered back down to the cliffs.

It was now mid-morning and, although the sun still hadn't broken through, it illuminated the cloud from behind like a startling white cinema screen. I walked northwards to Jubilee Corner, where I noticed a fresh landslide in the red topsoil, right next to the safety fences. It was clear that the soil was precariously perched on top of the chalk cliffs here, although I noted that four brazen guillemots had already moved in to make this freshly exposed ledge their home. They sat in a row, side by side, chatting away to each other. I was reminded of the famous photo of the construction of New York's Rockefeller Center, workmen sitting eating lunch on a steel girder, high up in the sky.

Back at Mosey Downgate, the guillemots, kittiwakes and razorbills were making a real racket. I met Kevin, an RSPB volunteer at the site. He wore his RSPB jacket and binoculars with pride and had a warm hat pulled close over his head. He pointed out the pair of puffins that I'd seen earlier. They hadn't moved but had caused quite a crowd of spectators to gather. He told me that the puffins had first been spotted on the sea at Bempton in late March and had started to gather in numbers before suddenly disappearing again. They only regrouped to come ashore at the start of April. They won't come ashore until conditions are perfect, and then do so en masse. 'Puffins usually nest in burrows in the soil,' he said, 'but the ground here is too unstable – they'd soon end up in the sea. So, at Bempton, puffins tend to find holes and crevices in the cliffs to use as nests. That's quite unusual.' He smiled. 'I love puffins,' he said. 'They're so clean. They have separate toilet areas, so they don't soil their nests. It's so civilised.'

I asked him whether there was any chance of further sightings of puffins, and where I could improve my chances. 'Puffins aren't stupid,' he said. 'Most will be down there' – he pointed towards the lower reaches of the cliffs – 'nearer the sea and sand eels.' I must have looked disappointed.

'But if you go and see Angela on the Grandstand, she's just spotted one.'

By the time the words had left his mouth and I had muttered a quick thank you – sorry, Kevin – I had turned on my heel and was doing my best speedy dad walk back to the Grandstand.

Like Kevin, Angela was also retired and chose to spend her free time being thrashed by cold winds on top of a cliff. She had white hair and the healthy glow of someone who spends a lot of time outside, together with glasses sensibly on a string around her neck. Thank goodness for RSPB volunteers, who do a sterling job. She ushered me to a nearby scope. 'There's a puffin in there,' she said. 'Just look down towards the left.'

I looked into the scope. There was wide crack in the chalk, on a slight angle. The chalk had greened slightly and at the base of the crack was a flat, sandy area perfectly protected from the elements. There was a puffin, I think, but it was asleep, with its back to me. Nothing but a black smudge that could have been a rook, or a small cat, for that matter. I told Angela. 'Yes,' she said patiently. 'He's asleep. We will just have to wait.'

I asked Angela about the current number of puffins at Bempton, especially after 2013 and the cappuffinccinos. 'Puffins have stabilised here, at around a thousand breeding pairs. That's really good. Bempton seems to be doing a lot better than other areas in England. I think it's all to do with the sand eels moving because of the warmer seas; puffins go where the sand eels go.'

We checked the scope. The puffin had gone. Angela

laughed. 'Maybe it's you, not the sand eels, that is the problem.'

I wandered off to the educational visitor centre to do a bit of research on climming, or egg-scooting, which was once all the rage at Bempton. In the 1800s, climmers descended the cliffs on ropes to collect seabirds' eggs, for use in everything from cooking to the shine on patent leather shoes. Apparently quite the spectacle, crowds would come to watch the men carry out their dangerous work. Climmers are said to have left every third egg to allow for the survival of puffins, guillemots and razorbills, or more likely to ensure that there were going to be eggs the following season. According to the RSPB, tens of thousands of seabird eggs were taken by the climmers at Bempton each year. Thankfully, the practice was stopped by the introduction of local by-laws and the Protection of Birds Act 1954, which outlawed the persecution of wild seabirds, including puffins. It struck me that it wasn't so long ago that the United Kingdom had a similar attitude as countries such as Iceland and the Faroe Islands, where seabirds and their eggs are still harvested.

After an hour, I was back with Angela. She saw me, smiled, and ushered me back to the scope, actually moving others to one side to make way for me. 'Have a look, but don't curse this one – it's my best,' she said, with a twinkle in her eye.

I pressed my eye against the cold metal of the scope. I looked into the crack. He was there. He was standing stock still, on his bright orange feet. There was no winter plumage on this one, no mud from fetching grass for nest building. This puffin was pristine. White-chested. Beautifully coloured beak. Smiling eyes. Pristine puffin.

I grinned and thanked Angela for this opportunity. She looked into the scope. 'He's gone! You've done it again!' she exclaimed, half-smiling. I made my apologies and left.

3

Puffin Surgery

April

Anglesey/Ynys Môn, Gwynedd, Wales

A well-dressed middle-aged lady presented the cashier of the gift shop at RSPB South Stack in Anglesey with a packet of biscuits. 'Are these suitable for coeliacs?' she asked.

'Sorry, I don't speak Welsh,' came the reply, and I had to leave, for fear my fits of laughter would see me knocking over the myriad puffin-themed trinkets and souvenirs. They did an excellent cream tea in the café though, which was packed full of families and birders, and where the staff were fully bilingual.

Ynys Môn or Anglesey is a flat, surprisingly green island at odds with the rugged North Wales coast from which it is separated only by the narrow but swirling Menai Strait. North Walians will laugh at you for visiting

Anglesey, saying the weather is always bad. As someone who holidayed in soggy Rhyl for the first sixteen years of his life, I find this somewhat ironic, reinforced by the fact that, like today, the weather is often brilliant compared to the mainland. Today the sun was shining, the sea was sparkling and from South Stack I could see the snow-capped mountains of Snowdonia and even the distant Llŷn Peninsula with its own puffin enclave, Bardsey, perched neatly on the end. My phone told me that I was in the Republic of Ireland and, presumably, started charging me accordingly. If I squinted, I could just about make out the coastline of that distant land, although perhaps that was my imagination. I might have been doing that for too long as I was getting weird-coloured shapes in my field of vision, so I decided to stop.

South Stack is on Holy Island in the north of Anglesey, near Holyhead. Holyhead is a bit run-down, and only a place to visit while awaiting the ferry to Ireland. It's a peculiar mix of old and new, but nothing really fits together and, oddly, you can still spot signs for the long-defunct Woolworths store, which reminds me of pick 'n' mix and cassette tapes.

The wonderful RSPB centre is at South Stack, where I found the café and shop. But it is Elin's Tower, a few metres away, that's the real heart of the action here. This squat, pristine, white building perched on the edge of a cliff high above the Irish Sea was once a summer house, built in the 1820s by a local family. It's quite a place to build a structure, perched on the cliff edge and looking more like a gatehouse to a castle. Inside, helpful RSPB staff assist visitors, while

CCTV images are beamed in from the cliffs. One camera was focused on a potential puffin-nesting spot; a local pair had been frequenting this crack in the rocks, sometimes with nesting material in their mouths. It was still early in the season though, and I could only see a pair of noisy razorbills in shot. A friendly female volunteer with a French accent told me that the puffins had been there twenty minutes ago, but I wasn't fooled. I'd played this game before. I've been to South Stack on quite a few occasions and, yes, it's true they have very few puffins – sadly now only twelve to fifteen individual birds for the whole season – but whatever time you arrive at Elin's Tower, the volunteers will always tell you that you've 'just' missed them. I know their game. I'm on to them.

I decided to walk back to the visitor centre, where I passed families picnicking on the grass and bikers arriving for a cup of tea and something greasy. Outside the centre is a three-foot-tall puffin statue. It's sadly been the only puffin I've seen on many of my trips here. I patted its cute, slightly weather-beaten, little head and hoped today would be different.

In the nearby car park, a chough was drawing a small crowd. Choughs are increasingly endangered members of the crow family. They have jet-black, *Game of Thrones*-style feathers, with bright red feet and a similarly coloured, slightly curved beak. Their feathers come down their red legs, like small pairs of black shorts. This one was happily searching out worms and grubs from around some marshy plants at the edge of the car park, living up to its Cornish name of palores, which aptly translates as 'digger'. I stood and watched for a while. The South Stack choughs were down to single figures at one point, but are now bouncing back. Numbers are increasing elsewhere in Wales too, including in Pembrokeshire and Snowdonia National Park. In Cornwall, a substantial amount of work has been done to help them return; the chough is, after all, on the Cornish coat of arms.

South Stack lighthouse sits on its own tiny island, separated by thirty metres of frothing, surging sea from the sheer cliffs that tower above. I made my way slowly down the four hundred steep and zigzagging steps. At the top, spring daffodils had given way to beautiful bluebells and bright yellow gorse was attracting early butterflies. The path was desperately steep and the steps oversized, but the views were enough to refresh a person. At every other turn, you could see the cliff face back towards Elin's Tower, and I inspected it carefully, hoping to find that the seabirds, and especially the puffins, were in residence. I had a good look, but I couldn't see them. I also kept a lookout for South Stack fleawort, a prettier plant than it sounds, with yellow, daisy-type flowers; it grows only here and nowhere else.

The angry sea was battering South Stack Island and, when I eventually reached the small metal bridge, I found it was all locked off behind a formidable metal gate and spikes to stop any unauthorised entry. I felt a little cheated. Someone had added googly eyes to the padlock though, which went some way to making me smile. Maybe it was too early in the year to access the lighthouse, or maybe it was because it was getting a fresh lick of brilliant white paint from a man with the longest roller in the world. It looked quite the job. Alongside the lighthouse were a long, low building, a series of whitewashed walls and several shags. Grass was growing where it could, but the lashing wind and sea ensured that most of the granite remained exposed.

On my way down, I'd passed patches of spring squill, baby blue in colour, so I sat on a bench pretending to admire them while getting my breath back. Punctuated by white scurvy grass, they were the true sign that summer was just around the corner. Like me, a common lizard was catching the early spring sunshine on a nearby rock. I decided that a return to Elin's Tower was my best bet, so I started the slow trudge back up the steps, complete with eye rolling and tutting from every person coming the other way as they shared the pain. There was some really intriguing lichen on the vertical rock, like the splodges of a tie-dyed T-shirt, but the puffins stayed away.

A huge rusting statue of a puffin looks out across the choppy Menai Strait from the former roadside Puffin café on the A55 North Wales expressway at Penmaenmawr. He looks forlornly and directly at Puffin Island. Although formally known as Priestholm, or Ynys Seiriol in Welsh, it is Puffin Island that has stuck. It's odd, because for a while there weren't any puffins on the island, or at least very few. Slowly and thankfully, they have started to return. I caught the boat from the pier at Beaumaris, name of which is taken from the French *beaux marais* – 'beautiful marshes.' It's difficult to see beauty now, with its squat, unfinished castle and hotels that have long since lost their glamour. A female busker was belting out *Crazy* by Gnarls

Barkley, slightly out of tune but with great enthusiasm. Even the local herring gulls were momentarily surprised enough to stop their own racket-making.

The pier was full of holidaymakers, most either waiting for boat trips or fishing for crabs by dangling multicoloured reels of line off the pier and hauling the hapless crabs in. This was likely the first time the creatures had ever been airborne, and they were now be fated to sit in a warm bucket of seawater next to the packet of bacon that had tempted them in the first place. The pier was fairly new and made of sheets of metal grille. I was made to wait at the top of the slope down to the dock. I had exchanged my ticket for a boarding card, presumably a vague attempt to make the trip feel more like an adventure. As a team of herring gulls mugged a bin for yesterday's sandwiches and soiled nappies, a child threw up chips over his dad's shoes and a girl walked past simultaneously eating two ice-creams. It couldn't have felt less of an adventure.

The *Island Princess* – a name more suited to a boat in Barbados or the Maldives, surely – arrived and we were herded like cattle down to the water's edge. I boarded the boat with around twenty other passengers and, oddly, two dogs. The boat had a jolly Liverpudlian captain at the helm, who told jokes over the tannoy. We set off, only to find we'd left four passengers ashore, so the *Island Princess* returned to collect them. The captain told the same jokes as we left for a second time. One of the dogs covered both ears with his paws.

It wasn't far to Puffin Island, and the views of North Wales and Snowdonia were fantastic. The views of Anglesey

46

were less so, seemingly comprised of ruined factories from the war, although I could be wrong as I was unable to work out quite what the captain was saying over the noise of the engines and the waves. Nearer to the boat was a line of coloured buoys, on one of which a juvenile herring gull was paddling fast as the buoy turned beneath him, circus style. He did well to keep his balance. The sea was calm, but grey, reflecting the overcast sky. I imagine the trip could be really special in the sunshine.

We passed the handsome black-and-white Trwyn Du Lighthouse, with its humbug design and strict 'No Passage Landward' instructions in capital letters. With the ferocious currents between Dinmor Point and Puffin Island, the lighthouse is very necessary. In July 2016, a motor cruiser subtly named *Le Babe* collided with the island and eventually sank, but not before the crew were rescued by a passing puffin-spotting boat and the RNLI attending from Beaumaris. Things can get choppy around there.

According to Natural Resources Wales, the number of puffins on Puffin Island was estimated to be around fifty thousand over a century ago but, by the 1990s, had fallen drastically to a mere twenty pairs. This was, in part, because of hunting (there are peculiar reports of puffins on the island being 'pickled' in wooden barrels), but was mainly due to the devastation wreaked by brown rats, which are thought to have made it ashore following a shipwreck in 1861. Rats scavenge the eggs and chicks of seabirds, causing a very real problem. It wasn't until 1998 that work started on ridding the island of its rats, with poison scattered by boat and helicopters from nearby RAF Valley, led by the

Countryside Council for Wales (now Natural Resources Wales) and the RSPB. The programme was completed in 2000, and no rats have been found since. One result of this extermination is that eider ducks and black guillemots have returned to the island, and it clearly has helped the puffin population. That, and less pickling, seem to have done the trick.

Past a large red buoy warning of a particularly nasty-looking rock, I spotted the cormorants first, tens of black dots amongst the thick grass. Puffin Island is a nationally important site for cormorants too. Some were alert on the light rocks by the shore, accompanied by the lighter-coloured shags, but the majority were up on the grassy slopes preparing their big, untidy nests. There were a few lesser black-backed gulls and herring gulls around, but on the south side of the island no puffins were to be seen. At the east end of the island, grey seals were lounging on the rocks. The captain pointed out a white, smaller seal cub, which had been born the previous September. There was a collective *aww* and the sound of hearts melting all around me. More seals followed – black, brown and grey, and some piebald, like cows – most of them dozing like lazy teenagers; the best they did was raise their heads as we passed.

As we rounded the corner to the north side of the island, the sea got choppier and several of the younger passengers started to cry as they were rocked from side to side with the motion of the waves. 'You'd pay £50 for this at Alton Towers,' the captain helpfully shouted. The limestone here was stacked like toast, only streaked with green algae and seaweed rather than butter. The birdlife on the cliffs changed from the gull and cormorant combo to pockets of guillemots and razorbills. This was more like it. A couple of oystercatchers picked at the seaweed on the rocks, my attention drawn to them by their bright orange feet, something that often gives puffins away too. I kept my eyes trained on the cliffs. There were not many seabirds there, but it was still early in the year.

'There's a puffin ahead,' tannoyed Captain Birdseye. Everyone craned their necks, moving to try to spot the bird. Another then flew directly overhead, clearly showing off. I got a great view of its colourful beak and desperately flapping wings. There was a smile on my face, and then another puffin was spotted in the water. This one was a real beauty, sitting on the navy-blue sea, peering over each wave as it crested. It had perfectly white cheeks and a lustrous, radiant beak like a miniature rainbow, making the yellow rosettes appear like the sun. It was a truly wonderful bird, and I was so glad to have seen it there, on its eponymous island.

A puffin had been found seriously injured on a beach in Anglesey and I found myself hopping into the car and driving to Stapeley Grange near Nantwich to meet it and learn about its recovery. Now an RSPCA wildlife centre, Stapeley Grange was once a stately home set among the green fields and footballers' homes of Cheshire. RSPCA-rescued creatures that require serious help often end up at Stapeley; thankfully, it gives the best care possible to our wild friends. Staff at Stapeley told me that they see a lot of foxes, badgers and bigger wildfowl, as well as birds from pigeons to buzzards, but rarely get asked to treat a puffin. In fact, this was only their eighth puffin in more than ten years.

Stapeley Grange is not open to the public and I was lucky to get access to see the team at work. I'm so glad that I did. I've never seen a group of people so dedicated to the welfare of animals.

On being found, the puffin was given – and I'll never understand this – chips and ketchup by members of the public, presumably until somebody, with enough sense to realise that puffins are not fans of traditional British seaside food, rang the RSPCA to report the downed bird. The attending inspector noted that the puffin had serious injuries to his wing and chest and immediately contacted veterinary staff at Stapeley to take him in. They obliged, of course.

The puffin was extremely weak, severely malnourished (despite the chips) and with necrotic tissue around the wound. Sara Shopland, the vet at Stapeley, explained to me that she X-rayed the puffin and then operated on it straight away, removing the necrotic – doesn't that word make your stomach turn every single time? – tissue and stitching up the wound. The wound was consistent with a bite, but she couldn't say whether from a dog or a fox.

When she showed me the X-rays of the puffin, I was astounded. I'd only ever seen X-rays of myself before and had mainly been either too squeamish to look or too confused by the blur of whites and blue-greys on black to take a proper interest. In Sara's puffin X-rays, which she'd taken from above and from the side, you could clearly see the bird's delicate bones and complex structure, as well as the surprising spread of its wings, which were like folded letter Ms on either side of its squat body. Sadly, you could all too obviously also see the damage caused by the attack, and exactly the task that Sara had faced.

She persevered though. The operation went well, and the little bird showed remarkable fighting spirit. Sara prescribed a daily diet of whitebait and sprats, as well as antibiotics and vitamins, to give him the very best chance of survival. And survive he did, putting on weight and regaining his strength. On arrival, he'd weighed only 320 grams, well below the ideal 500 grams of a healthy adult. At the time of my visit, he was 460 grams.

I was ushered into a treatment room, complete with medical equipment laid out on shiny, ultra-clean stainless steel, a folding examination couch and those concertina curtains that you don't see anywhere else. I've seen doctors' consulting rooms less well equipped. It was made very clear to me – and quite rightly so – that this wasn't some sort of parade or zoo-style viewing for me. The puffin was receiving his daily treatment, and I just happened to be there. The welfare of the bird was the absolute priority here, and nothing was going to compromise this.

One of Sara's colleagues disappeared for a few seconds and then reappeared with a folded blue towel under her arm. From under the towel there slowly emerged the puffin. He was calm and not at all distressed. I was thrilled to see this enigmatic little bird. He was inquisitive and tilted his head from side to side in that peculiar puffin way. His beak was still orange and yellow, although slightly faded and dull – the way a human's eyes go after a spell in hospital. Sara and her colleague handled him confidently and professionally.

I saw a gap in the feathers under his right wing and some slightly red scarring. It was all healed though and looked clean and healthy, if not a little pink.

Sara showed me some stress bars on the puffin's wing, caused by the trauma, before cleaning the site again and applying some moisturising lotion. If I was concerned about the puffin's recovery, he had his own way of telling me that he was doing just fine. He let out a little puffin caw and then dropped a splodge of thick white puffin poo onto the pristine floor. Sara, clearly well practised in such matters, moved her feet quickly and deftly in the manner of a professional dancer, to avoid a splatter of the landing emulsion.

The puffin was doing well, but clearly wasn't quite right. Even I, with my limited veterinary skills, could see that. In addition to his dulled beak, the usual bright white plumage around the eyes was grey and darkened. I suspected that this was partly a change into winter colours but was also no doubt a sign of how much recovery was still ahead.

This was not, of course, the first time a puffin had been found and rescued. In fact, in spring 2011, a puffin was found in fairly unusual circumstances – at a sexual health clinic in Hampshire. No, really. The poor puffin, a juvenile, was found in a basement building at Winchester's Royal Hampshire County Hospital. It didn't make it.

I stayed in touch with Sara throughout the puffin's recovery. She very kindly kept me updated at every stage: the wound healing up, the feathers growing and becoming stronger, the release into an outdoor setting, and his successful return to swimming in the Stapeley pond.

In November, I received a message that the puffin was to be released. The centre had done some research, identified a suitable location on Anglesey and found a day with perfect weather conditions. After careful deliberation, they had decided that this was the best thing for the welfare of the puffin, aided by the fact that there were several little auks in the area – the little auk being a relative of the puffin. RSPCA staff escorted him from Cheshire to Anglesey before releasing him on a shale beach. They were good enough to send me a photo of him standing on the beach, staring apprehensively out to sea. The last they saw of him, he was heading out to sea. All we can hope is that after Sara's and the team's hard work, and the fighting spirit of that little bird, that he survived and has done OK. He'd certainly been through the works that year and deserved a little break. I'd hate to think that he didn't make it and was found some months later outside a fish and chip shop in Beaumaris, obese and lazy from a diet of greasy chips and bright blue fizzy pop.

4

Little Welsh Pals

May

Skomer/Ynys Sgomer, Pembrokeshire, Wales

Martin's Haven sits on the very point of the very tip of the Dale Peninsula in Pembrokeshire, South Wales, and is only accessible through narrow, twisting lanes with thick green hedges sparkling with yellow and pink flashes. This little hideaway is a slice of heaven, a tiny pebbled cove that meets the Irish Sea between two grassy knolls. Tiny coloured boats bobbed on the twinkle-specked water and an old winch lay on the concrete slipway, rusting into the ground. There was a tiny display of information boards, mainly on wildlife, some rather run-down toilets in the converted fishermen's cottages, and an elderly-looking piece of whale vertebrae. The white-painted Lockley Lodge, named in honour of RM Lockley, more of whom

54

later, sold everything puffin-related you could think of, from T-shirts to tea cosies, as well as hot coffee for cold days and the best, creamiest Welsh ice cream for the hot days. Outside, a house martin was collecting mud to start building, its head as shiny and greasy as a 1950s gangster's hair.

It was the start of May and the hedges were just sprouting into life, catching up with the sticky weed favoured by schoolboys the country over. The sea was so clear and so inviting that you immediately wanted to strip off and dive in. If you did, though, you'd run back up the pebbled beach shrieking and with the first signs of hypothermia from the intense cold of that deceptive water. Take this from a man who knows, and who retained little dignity. Also understand that this is exactly why the divers that flock there wear thick dry suits as they waddle awkwardly down the path to the water's edge – they can't resist the clear, beautiful waters of the small bay either, but come fully prepared. Sensible.

Everyone knows that this is the place to catch the boat to Skomer. Most of all, the people running the trips know this. The first boat for which I attempted to buy a ticket was sold out. It wasn't leaving for over an hour, but all fifty tickets were gone. I managed to get a spot on the second trip of the day, at ten thirty. This left me an hour to drink coffee and admire my surroundings before I was herded on to the *Dale Princess* with forty-nine other Skomer goers. There was less of a holiday mood than aboard Anglesey's *Island Princess*; the Skomer passengers really cared for the wildlife they were hoping to see. The sea was gentle, with the odd cresting wave. The sky was blue and streaked with contrails. As the crossing commenced, spray caused some passengers to gasp with surprise at the cold; others hoisted oversized telescopic lenses into the sky to photograph the gannets that were following us, presumably taking the fifty humans squeezed into the tiny boat as sardines of a different sort. The sun was so glaringly intense, I was surprised the photographers didn't burn their own retinas out.

On landing at North Haven, we were ushered onto a makeshift harbourside that led up to a set of large steps – the sort that are too high for one step but too shallow for two. This luxury soon ran out though, leaving nothing but a dusty track. We were greeted by a late-teenage male volunteer guide, who was tanned and wearing shorts, clearly with thoughts of last summer's gap year in Sydney. He gave the lowdown on the island before being impressively interrupted by a soaring peregrine, which appeared directly above him and immediately and fully took the attention of all fifty new arrivals. The peregrine soared effortlessly on the thermals. I may have imagined it, but I fancied I could see its dark moustache too. It was soon joined by a second peregrine, and the air cleared of all other birds. The guide gave up trying to talk for a while as we gazed in awe at these two beautiful birds. Unfortunately, neither bird did one of those zero-gravity, terrifying, freefall dives that they use to take out their prey. We were lucky to see them though – one of only 1,500 pairs in the UK, following a sustained period of illegal persecution, mainly to protect gamebirds, that is sadly ongoing.

The guide regained our attention. He told us to stay on the path and away from the burrows, for fear of collapsing them and crushing the resident puffins or Manx shearwaters. He told us all the things we could not do. Don't leave litter. Don't go too near the cliff edge. Don't try to catch puffins and put them in your rucksack. The wind was blowing dust into my ears and eyes, and the lecture, as necessary as it was, began to grate, so I was pleased when he finally gave up and let us go on our way.

I continued up the rough track and headed straight out to the Wick. My previous knowledge and the power of Google told me that this was the best place on Skomer to see puffins. Puffin in Welsh is pâl, by the way. How nice is that? Puffins are your little Welsh pals. I began wondering if puffins had different accents depending on where they returned to breed, and vowed to find an expert to ask. In the meantime, I amused myself by imagining a little puffin called Pal spending his time speaking Welsh while living on a remote Welsh sea stack.

Once at the top of the slope, I was rewarded by the most astonishing view. The island was not all pale green, as I was expecting. The entire top of Skomer was carpeted in the deep purple of springtime bluebells. They were everywhere, fighting the sky for primacy of colour. There were gentle outcrops of grey rocks, and white dots of lesser black-backed gulls stationed around. The purple though, the endless, luxurious purple, was breath-taking.

On the blustery traipse to the Wick, I crossed a shallow valley with a small stream running through it. This is known as South Stream Valley – I know; the imaginative name is too much – and it too was purplish blue with bluebells, looking down the valley to the sea and beyond to the Neck.

The bracken had yet to return to the valley for spring, but there was some red campion in the grass like twinkling pink stars at my feet. Nearby, a blackcap was perched on top of a barren branch that gently swayed in the wind. I could see why he'd chosen this spot; it had a view and the sun was warming both of our backs.

The Wick already had several visitors, cameras in hands, clicking away. As I approached, it was not birdsong I heard but the beeping and clicking of smartphones and DSLR cameras. The closer I got, the easier it was to see why. There were puffins everywhere.

The Wick is a sharp inlet of cliffs between the main part of Skomer and South Plateau. It is very familiar to me, as a framed painting of the Wick – a gift from a friend – hangs in my living room. The south cliff is sheer and home to lines of razorbills and guillemots, while the opposite side is less steep and occupied by a variety of gulls, whose filthy, paint-like smudges are all over the place. The point where the two cliffs meet was puffin central. The puffins used this place like a busy railway concourse. The path ran parallel to the clifftop, delineated by a flimsy rope fence that was presumably more a symbol than a safety measure. The path was worn from both human and puffin use, and the orangey soil showed through, blowing up into our faces whenever the wind was particularly strong.

Puffins strutted up and down just in front of the fence, parading around like supermodels. They preened themselves and did their adorable tail shake before flapping their wings and posing for more photos. Preen. Tail shake. Flap. Pose. Repeat. They appeared almost fearless of us humans. There are a couple of theories about this. One is that they have simply become used to our presence and tolerate being photographed at every opportunity. The other is that they are using us; in other words, they realise that with plenty of Gortex-clad humans about, they are fairly safe from predators such as peregrines and great black-backed gulls.

Either way, the puffins appeared to be unafraid and continued semi-running back and forth across the path to their burrows. Studies have shown that puffins tend to have two types of walk. One is where they are stood proud, often near a mate or a burrow with head held up high and chest pushed forward. The other is their low-profile walk, designed to get them back to the burrow inconspicuously. This walk reminds me of commuters on London's Circle Line – head slightly lowered, no eye contact, a hint of hurry.

We were so close that, while remaining firmly on the path, I was able to peer down burrows to search for puffins. This was the bit that I really loved. Frequently, I was met with a white face and intense eyes peering back at me. If I waited a while, puffins would emerge from their burrows with slight trepidation but with a frequency that put me in mind of the arcade games where you're encouraged to whack creatures with a foam hammer. Not that I could ever whack one of these fellas – they are way too special, and protected.

While gazing into the orangey-brown earth, I was struck by the sheer beauty of the puffin's face. The slightly sad eyes and the bright yellow dandelion crowns either side of the beak. The beak itself is so much more than you see on photographs; it's orange like an organic egg yolk, warm and rich, and with slices of lemon yellow and then dark, night-time blue. You'd have trouble rendering such colours in paint or any other material. It really is truly remarkable.

Two little girls, no older than five, ran from burrow to burrow, shouting, 'Let's find another one,' and then squealing with joy when they saw a puffin emerging, blinking into the bright sunlight. I was tempted to join them; I felt the same level of excitement.

There is growing evidence that puffins are a lot smarter than we once thought. A study by Dr Annette Fayet from the University of Oxford on Skomer, and replicated on Grímsey in northern Iceland, has shown that puffins can use tools. Puffins on both islands were seen to use sticks to scratch their backs and chests. She concluded that the use of such tools for self-care could indicate a higher level of intelligence than we originally thought.

We already knew that puffins are smart because of the way they construct their burrows. Right on cue, I spotted a puffin still preparing their burrow. He scooted down the earthen hole, only to back out slowly, using his sharp claws to propel soil backwards out of the hole. Burrows are a metre to a metre and a half long, with a bend partway to deter unwelcome visitors. Puffin burrows, as Kevin in Bempton informed me, contain separate chambers for the bathroom and bedroom. The pufflings need to be clean and have feathers that aren't soiled to give them the best chance when heading out to sea, so a 'no soiling the bedroom' policy makes sense. Rules to live your life by. The burrow architecture is cleverly designed with a slight gradient to keep moisture away from the egg and puffling. The puffin demonstrated an effective if rather messy way of clearing and widening the hole, and returned to the surface looking pretty pleased with himself, despite the substantial amount of mud on his white bib. I guessed that was nothing a dip in the sea wouldn't sort out, and he had the same idea, heading over the path and under the rope before gracefully leaping from the cliff, orange feet pointing outwards, towards the waiting sea.

I took a seat on a nearby wooden bench and broke out my picnic of soggy sandwiches, broken crisps and squishy chocolate while waiting for the crowds of people, not puffins, to thin out. Most visitors seemed to walk the circumference of Skomer in a clockwise direction, but I was content to stay exactly where I was. Once the crowds had gone, I settled down to some proper puffin watching. Despite the gusting wind, it was bliss. The puffins were completely at ease. One was less than a metre away from me and, for about fifteen minutes, I watched her preening herself and flapping and stretching her wings. Then she shook her tail feathers from side to side. I was transfixed. I could see that the black feathers on her back weren't completely black but were flecked with white and had a delicate sheen too. She shook her tail feathers again, before scooting backwards in a peculiar manner. She then raised her bottom slightly before shooting several streams of fishy liquid yellow excrement towards me. I was most surprised, if not a little put off my cheese and piccalilli sandwiches.

Taking the hint, I decided to up sticks. I made my way back down the path to the remains of an old Roman fort and then across a low boggy area known as Moory Mere.

There had been sightings of a woodchat shrike in this area in recent days, so I kept my eyes open. Shrikes occasionally have GPS hiccups and overshoot during migration, making a rare but welcome visit to the UK. I didn't see the bird, but I did see several men in hides telling each other stories of better birding days. At one point I fancied that I caught a fleeting glimpse of the shrike but, thinking back, it may have been merely a wheatear at speed.

I headed into the centre of the island, to an area known as Old Farm. Old Farm has been pressed into use as accommodation for holidaymakers and staff and also contains a rather damp and disappointing 'visitor centre' and some slightly smelly long-drop composting toilets. Near the farmhouse, several families were having lunch while pouring over guidebooks and trying to decide on their next move, including the best puffin-spotting sites.

Around the farm there was plenty of evidence of another burrowing creature. The good old-fashioned bunny. On Skomer, rabbits are prolific, leading to some areas being fenced off to protect rare plant species from the marauding fluffy-tailed creatures. The rabbits at the farm were obviously not afraid of humans – quite a few were hopping happily around in the undergrowth. It's worth checking before you throw your picnic rug out on the ground – those are not chocolate raisins.

Having rabbits on Skomer actually helps the puffins. Not only do puffins sometimes appropriate rabbit burrows for their own use, but the rabbit's constant grazing keeps the vegetation down, in line with puffin preferences. I also saw plenty of evidence that the rabbits are a food source for peregrines, owls and great black-backed gulls. There were quite a few rabbit carcasses around the farm and, on the path, there was a particularly grisly specimen – a furry skull with cartoon-like teeth, exposed vertebrae and no sign of limbs. It had dried out in the sun and seriously quelled my appetite as I headed back up to the Wick.

It's not only desiccated rabbit carcasses that are frequently spotted on Skomer; Manx shearwater carcasses are an all too familiar sight across the island too. This is not so surprising, as Skomer is one of the most important Manx shearwater breeding grounds in the world.

The Manx shearwater might just be the original puffin. Its Latin name, *Puffinus puffinus*, was somehow stolen and Anglicised by the puffin itself, leaving the Manx shearwater to take its name from the Isle of Man where it once bred. Although slightly larger, Manx shearwaters share a number of similarities with puffins, which may be where the confusion arose. Both species lay a single egg, for example, and make use of cliff top burrows in similar locations. Both are mainly black and white – a design that makes them more difficult for potential predators to spot on the sea from above or below. In centuries past, both birds were a staple source of food, and even acted as currency. That's where perhaps the similarities end. Manx shearwaters migrate huge distances over the ocean to South America using their powerful, extra-long wings. When it comes to breeding, they also prefer to operate under the cover of darkness and will wait silently on the sea in large rafts until darkness falls. Using their excellent eyesight, they make their way to land, before letting rip with an ungodly, eerie wail that has had sailors believing that entire islands were inhabited by tormented trolls. It sounds something akin to a duck being slowly but determinedly strangled. Manx shearwaters struggle to walk at all well on land. This makes them easy prey; it's tough work being a Manx shearwater, as the quantity of half-eaten remains on Skomer will attest.

I resisted the temptation to return to my usual seat at the Wick and, instead, headed on to Skomer Head, with frequent sighting of puffins on the clifftop as I went. The wind really had found its force now, and dust blew over the top of my sunglasses and directly into my eyes. It stung. From Skomer Head, and between clouds of dust, I could see Skokholm – Skomer's slightly smaller neighbour. You can stay on Skokholm but, unlike Skomer, it doesn't allow day-trippers. I promised myself that one day I'd make it there and stay for a while, if only to go off-grid and spend some time with only seabirds for company.

Back at the Wick, I found a seat and pulled a battered copy of *Puffins* by RM Lockley from my bag. I really cannot go without mentioning this chap when writing about Skokholm and Skomer.

Ronald Lockley fell in love with nature as a child and devoted his life to nature and conservation. After moving to Skokholm in his early 20s, he became a nature writer before that profession was fashionable, writing more than sixty books, covering not only the birds and mammals that he found around him but the locations and characters too. Lockley turned Skokholm into the first bird observatory in the UK, and fought hard to protect his adopted Welsh islands, including against the proposal of an oil refinery off

Milford Haven. Later oil spillage disasters may just have proven him right. In later life, he moved to New Zealand, where he continued to write and advise the New Zealand government on conservation matters.

Lockley lived to the grand old age of ninety-six, a true pioneer and, undoubtedly, the reason I was able to sit on Skomer and enjoy the birds, especially his beloved puffins. Fittingly, his ashes were returned to Wales from New Zealand and scattered on Skokholm.

Puffins was published in 1953 and is a great little book if you ever get the chance to lay your hands on a copy. Lockley writes in a gentle, easy-going style and there are wonderful pencil sketches by Nancy Catford of smiling puffins and cheeky rabbits. One of Lockley's books on rabbits is said to have heavily influenced Richard Adams's bunny-based novel *Watership Down*, and it is easy to see that influence in the characters that Lockley gives his subjects. Lockley spends a great deal of his book following a puffin pair on Skomer through the breeding season, affectionately naming them Frater and Cula after the puffin's Latin name. His reports are frequently amusing, but show the intensity with which he studied the birds: 'Frater, like every other puffin in view, watches each puffin near him as it alights or takes off, or does anything active. He cocks his head sideways to get a better view of each performance. Query: If he cocks his head sideways, he will see the performance with one eye only; then what is the eye on the other side of his head seeing? Does he 'think' with the outward turned eye only, the other being turned inwards and temporarily unfocussed?'

In relation to the endearing relationship between Frater and Cula, Lockley reports, 'Frater reappears and stands on his lawn looking round as if for Cula. The chick, far inside the burrow, is silent. Evidently the feed of ten fishes it received and possibly another meal in my absence has satisfied it. Frater at least has evidently no intention of collecting more food for it today, and beyond an occasional visit he is content to idle.'

The sun was shining at the Wick and I had endless opportunities to watch the antics of the Skomer puffins while dipping into Lockley's thoughts. He'd be delighted to hear how well puffins are now doing on Skomer. Just like Frater, I was content to idle there too.

5

Puffin Festival

May

Coquet Island and Amble, Northumberland, England

Amble is a wonderful little town south of Berwick-upon-Tweed on the rugged Northumberland coast. It has quirky harbourside shops in wooden pods, some restaurants that are hotly tipped (by me) for a Michelin star – Sea and Soil, I'm looking at you – and new apartment buildings along the quay. Occasionally, Amble is bafflingly out of date – see the DVD rental shop on the High Street – but it is outstanding for one week of the year, when it celebrates everything puffin with a Puffin Festival and, during my visit, puffin signs adorned the routes into town and on every lamppost to prove it.

The town was thronging with people of all ages, and the central square was full to the brim. There were stalls selling puffin-shaped crafts – I bought a rather cute but slightly overpriced purple felt puffin, which I have since inexplicably lost – fairground rides, various wildlife organisations touting for business, and a hotdog stand that filled the entire place with the smell of fried onions. People were happy in the sunshine, hotdog in hand, while the kids played around their feet.

The biggest attraction, though, was a human in a puffin suit. He walked around with his minder, or guide, however you wished to see it, and went by the name of Tommy Noddy, a tribute to the Scottish colloquial term for puffin. Tommy had an impressive beak on him and two slightly chunky orange legs stuck into massive feet. He wandered around the square like some kind of celebrity, hugging children, posing for selfies and flapping his oversized wings. I couldn't resist, of course, and found myself posing for a photo with Tommy just like everyone else.

The Puffin Festival has been held in Amble every May since 2012 and is timed to coincide with the best time to see puffins on nearby Coquet Island, which lies just a mile off the coast. It is, of course, also designed to promote the town's independent shops, nearby Queen Street being chock-full of them. From what I saw, it seemed to be working. There were various activities throughout the week, including talks, kite flying, surfing lessons, guided nature walks, local history talks, photo exhibitions, live music and even a fringe food festival. An impressive effort.

It strikes me that the draw of the puffin is something special. You don't, for example, get a festival celebrating the great skua, or cartoons about a Manx shearwater, for that matter. I understand that the puffin is colourful and appealing to look at, but I think the fascination goes deeper than that. It seems to me that the intrigue around puffins is more than merely superficial, albeit one that's good for tourism and trade.

It was a pleasure to feel the sun on my arms as I sat on a bench in front of the small, quaint harbour and watched a family from Japan narrowly escape an aerial bombing from a team of black-headed gulls. I actually didn't see any herring gulls in Amble; it seemed that the black-headed crew had taken over the town and were surviving mainly on chips and ice cream, which probably explained their loose bowels. I didn't attempt to explain this to the Japanese family, who had retreated to the nearby RNLI building for safety.

I wandered around the town – also known as 'the friendliest port' and, my favourite, 'Amble by the Sea'. After a hearty pot of local smoked fish chowder and crusty bread, followed by an abnormally large slice of white chocolate cheesecake at the Pride of Northumberland, I found myself back at the harbour feeling slightly podged. This is a real working harbour. Primary-coloured fishing boats sat on the wet sand waiting for the tide to return, a busy forklift truck lifted crates of fish into refrigerated lorries, and blue-and-green lobster pots were stacked everywhere. On peering into one, I saw a desiccated crab that had fallen for the lobster pots' charms but had clearly been too stubborn to release itself from the coarse blue rope of its prison when its mates were tipped into a waiting fish box.

A faded sign near the harbour, its paint peeling from years in the salty air and rain, advertised puffin tours to Coquet Island. There was a handwritten mobile number on it, so I gave it a call. A Geordie voice told me that the next boat wasn't until 6pm and, in any case, it was fully booked. My heart sank. The Puffin Festival was evidently too successful – I couldn't possibly visit a puffin festival and not see a single puffin. And, no, Tommy didn't count.

To try and lose a few calories from the cheesecake that was now resting heavily on my internal organs, I followed the wooden boardwalk to the breakwater that led to the red-and-white striped lighthouse. The lighthouse is best described as small but eye-catching, although I couldn't get any nearer due to some frankly over-the-top fencing and menacing red signage. The local lads were fishing from the seawall nearby, and there was a view to the tantalisingly close Coquet Island across the smooth sea.

Coquet Island is relatively small compared to other islands I've been to. It would be hard to spot in foggy conditions, or at night, and I guess that's a very good couple of reasons why it too has a lighthouse. The island is owned by the Duke of Northumberland and is uninhabited save for a handful of hardy RSPB wardens during seabird season. For over fifty years, the RSPB has protected puffins, terns, eiders, kittiwakes and Mediterranean gulls here. The duke doesn't seem to mind sharing his land with around 40,000 seabirds. It must be a desolate, empty place in winter once the birds and wardens have departed.

There is one very good reason why boats are not permitted to land on the island. Actually, there are now quite a few of these reasons – the endangered roseate tern. In the UK, their numbers once dropped to a mere eighteen pairs. They take their 'roseate' name from the wonderful pinkish blush that adult birds display on their otherwise white underbellies during the summer, something that sadly made them attractive to the millenary trade in the nineteenth century. Roseate terns are poised and charming birds, with a handsome black cap and long, elegant tail streamers like a child's kite. They return to Coquet Island each summer to breed – the only place in the UK where they now do so – and are welcomed by RSPB volunteers who set up bespoke nesting boxes for the terns, remove rampant vegetation that may impede them, and monitor nesting sites using a complex system of lasers and mobile phones to ward off predators and even egg collectors.

I've never understood this awful, despicable hobby. Why would a person steal something so rare and, by doing so, ensure its death, just for the chance to add to a collection of other cold, dead things? Why not allow it to develop into something truly beautiful? It's not just volunteers and wardens who are helping to the birds here; in 2017, prisoners from HMP Northumberland made and donated twenty-five tern rafts for use on the island. It would be a delicious irony if at least one of the prisoners was serving time for egg collecting.

I'm pleased to say that the conservation work by the RSPB on Coquet Island has started to pay off, with an increase to a fantastic 130 pairs of roseate terns in 2020. The best thing is that you can watch the whole thing from the comfort of your home – CCTV has made the roseate terns of Coquet Island into YouTube stars.

I'd dawdled too long and was now late for my next event at the Puffin Festival. I ended up doing an ungainly half-run down Queen Street to the large gift shop where Jenny Colgan was giving a talk. Jenny is an accomplished and bestselling author who writes in a variety of genres including romance and sci-fi. She's even a writer for *Doctor Who*. I hadn't come to hear her talk about sonic screwdrivers though. Today, she was talking about another love of her life – puffins. Jenny writes about a puffin called Neil, and his saviour, a young girl called Polly. I won't spoil her stories here, save to say that they are adorable – even more so once you realise that she's also responsible for the cute illustrations of Neil – and that Neil only ever says 'Eeep!'

Children were sitting on cushions at Jenny's feet and the adults formed a semi-circle behind them as Jenny spoke in her wonderful soft Scottish accent. It turned out that her children were also in the audience and, as she started to speak, she had to shush them – they quite clearly had already heard the story of how Neil came about. In fact, Polly – the young girl in her tales – is based on Jenny's daughter. Jenny explained that, like Polly, when they lived in France, her daughter had a puffin toy as a *doudou* – a security blanket. She took it everywhere, and it was as filthy as it was loved. One day, the *doudou* was lost, causing much heartache. Jenny launched a social-media campaign to try and locate the lost toy and, despite shares and likes across the globe, the toy could not be found. There were some heroes in this piece though – Amble Puffin Festival quickly dispatched a replacement puffin. How lovely is that? I could see why Jenny had been so keen to repay the favour.

Jenny revealed that she had to cut her talk short so that she could join the 6pm boat trip out to Coquet Island. Most of the audience appreciated this as they were going on the trip too. But I was seething. Not with Jenny, but because I hadn't been able to get on the boat. I hid my feelings and went to meet Jenny. She smiled at me, and even drew a Neil Puffin in my copy of her book. 'Eeep!' I said. Not really.

I left the shop and skulked back to the harbour area, where I treated myself to fish and chips and sat on a rickety wooden picnic bench to eat. Fish and chips may always taste better at the seaside, but I can never get on with those cheap wooden forks. I ended up messily shovelling flakes of succulent cod into my face with my fingers. It wasn't pretty, but the only ones watching were a menacing crew of twenty black-headed gulls awaiting my scraps. They did not know me well.

As I devoured my meal, a queue began to form for the puffin tour out to Coquet Island. My heart sank all over again. I was staying some forty miles from Amble, so a return journey would not be so easy, plus sailing conditions today were perfect. The deep blue sea was as still as a millpond, and the sky was so clear it was difficult to tell where the one ended and the other began. It was also the Puffin Festival. I couldn't possibly come to a puffin festival and not see a single puffin, surely? That was not what *Every Last Puffin* was about. No, a real puffin was required.

To add insult to injury, I spotted Jenny in the queue with her kids. She was even wearing a nautical-style striped jumper and deck shoes. In all, about twenty eager puffin-spotters boarded the white-and-navy-blue boat, which had slowly chugged into the tiny harbour and moored alongside the green concrete steps. I finished up my chips, much to the disdain of the gulls, and set off to find a bin.

As I did, I noticed that a second boat had taken the place of the now departed tour boat. This was an impressive craft, a former RNLI lifeboat, still with its bright orange livery and the red stripe on its bow. I later found out that it was a restored forty-one-foot lifeboat that had previously worked the wild seas off Eastbourne. Now known as the *Steadfast*, it was certainly eye-catching, but not as eye-catching as the sign on its side, which said 'Puffin Cruises'.

I shoved the greasy chip papers into an overflowing bin and hotfooted it to the top of the steps. I was met by Dave, a retired mental-health nurse from Newcastle. He was short, with a round, smiley face, and extremely friendly. He told me that there was plenty of space on the trip and that their time of departure was in five – or six – minutes. I was invited on board and took a starboard seat on the carpets that had been placed on the boat's benches. I smiled to myself and felt extremely lucky. I was going to see puffins after all, although I was unsure whether five – or six – minutes would be enough time for my fish-and-chip supper to settle, and I feared that the fish might be unexpectedly returned to the sea.

Dave was decidedly chatty. He told me why he was no longer a mental-health nurse – due to a cancer scare – and that conducting puffin tours, especially in the summer, had given him a new lease of life. He promised to keep me informed and said that if he didn't know the answers to my questions, he'd just make them up. He smiled. The captain was Dickie, who stayed firmly in the orange cabin of the *Steadfast*, which was adorned with stuffed (toy) puffins and a pirate teddy bear sporting an eye patch.

With the boat half full of passengers, we gave chase to Jenny's boat. Slowly. Calmly. I couldn't imagine Dickie and Dave doing it any other way. It was a wonderful evening and the short trip out to Coquet was made much easier by the gentle evening sun warming my face and arms, and the light swell rocking the boat softly. If I hadn't been so excited about the puffins, I could have easily fallen to sleep.

We approached Coquet Island from North Steel, a flattish, tidal area clearly favoured by common seals. There were a good number on the rocky beach, some of whom gently raised their heads as we passed. The lighthouse and former monastery were now visible, glowing white in the evening sun.

The puffins were not hard to spot. I had been concerned that not being allowed to land would mean a limited chance to see the little critters, but I needn't have worried. They were all around us. The sky was full of puffins to-ing and fro-ing from the island, and the sea was speckled with floating puffins diving for fish. As the *Steadfast* approached the seaborne puffins, some tried to fly away, flapping their short wings frantically but only just skimming the waves, while the smarter ones merely lowered their beaks and disappeared beneath the shimmering surface.

We continued around the coast at a sedate pace, all eyes on the island, which was topped with a green crown. On a sandy outcrop, I saw the square wooden nesting boxes that had been placed there by the RSPB to help the roseate terns. Dave noticed my interest. 'What they didn't realise is that the puffins arrive back first, before the terns. The puffins moved into the boxes and had to be moved out again when the terns arrived.' The thought of wily puffins taking advantage of a readymade starter home made me smile.

He told me that he'd known of up to forty thousand puffins being on the island, but this year there seemed to be only around twenty thousand. He blamed this on the lack of sand eels in the area, but hoped that next year would be better.

Once we'd rounded South Steel, there was a slightly raised area, not what I'd call a cliff, as it was only a few metres high. It was vertical though and, on top, I saw hundreds of puffins lined up. It was like they were at a rock concert, waiting for the headliner to appear. Occasionally, a single puffin would leave the capacity crowd and fly out to sea, passing over our boat. Their orange beaks were the exact same shade as the *Steadfast* cabin, and their white underbellies glowed amber in the setting sun. Outstretched feet, ready for landing, became miniature stained-glass windows.

I was absolutely enthralled by the spectacle of the puffins. Dave appeared beside me. 'Aye, it's a new lease of life,' he said. I wasn't sure whether he was talking about himself or the puffins.

6

Puffin 1: Gull 0

May

Farne Islands, Northumberland, England

'Ooooch!' I shouted, and instinctively put a hand up to my scalp. It was covered in blood. I had been too busy watching the others and using a technique I'd learnt in Iceland – holding one hand up in the air to present a higher point – to realise that I was also under attack. I felt another sharp jab in the back of my head that pierced the skin and sounded like it had connected with my skull. I needed to hurry on. The technique wasn't working.

Arctic, sandwich, little and common terns all nest on the Farnes. It's the Arctic terns that take time out of their day to meet and greet new visitors to the islands though. As soon as *Serenity* moored up to the newish-looking concrete jetty on Inner Farne and visitors started walking up the gently sloping path, they started.

Like the roseate terns on Coquet Island, Arctic terns don't really nest but make very shallow scrapes on the land into which they lay their precious green spotted eggs. It's a strange strategy and one which requires much defensive work to keep both eggs and delicate chicks safe from predators. Actually, it's less defensive and more of a full-on aerial attack as, with a loud screech, they plunge beak first into any potential predator within a few metres of their scrape, and they are not afraid to draw blood with their savage, knitting-needle-sharp beaks. Their beaks are bright red; it's not clear whether they are naturally this colour or if this is the stain of human blood.

The entire line of passengers was now being attacked by a squadron of terns, making people dive for cover, hold items over their heads, and adopt a peculiar bent-at-the-waist-and-knees walk that John Cleese would have been proud of.

It was late May, and the day had started at the deliciously named Seahouses on the Northumberland coast, where the Farnes had manifested as several dots amidst a thick blanket of sea fog. The sea was calm though, and the kiosks near the pier were doing a brisk trade in tickets out to the islands. I don't suppose that the Farnes' appearance on the BBC One *Springwatch* programme had done anything to harm sales. Puffins had featured heavily and I had been told that the population was faring well. I couldn't wait to find out.

I purchased tickets from a friendly lady with a Geordie accent and followed the crowd down to the jetty to wait for the boat. Appropriately, it was called the *Serenity*. I hopped on board and secured myself a seat. Searching for puffins had taught me this if nothing else – the British public shows no mercy in grabbing the very best seats on boat trips. *Serenity* was a newish-looking pristine white catamaran with stainless-steel fixtures and fittings. The captain looked rightly proud of his vessel and welcomed us aboard with a warm smile and a handshake.

He took us past Inner Farne and across Staple Sound. Like belly buttons, the Farne islands are distinguished by their position, Inner or Outer, depending on which side of the Staple Sound they lie. He was heading to the Pinnacles, on the south point of Staple Island, which rose out of the grey sea against the grey sky. The sea fog had lifted, and I found the soft greys utterly beautiful, like being wrapped in cashmere. This did not come across in photographs, as devices can't distinguish between the soft hues, rendering them all into a single block of drab colour.

The Pinnacles are an impressive series of sea cliffs striped with different pastel shades. The bottom was a light, sandy brown, from the tides keeping the rocks clean and bare; the weather had sent the next layer into a slate grey, and the top layer was a temporary, but garish, white, the result of a healthy population of seabirds. And there were plenty there. The cliffs buzzed with cormorants, shags, kittiwakes and guillemots. Especially guillemots, which filled nearly every available space, nestled dangerously close to one another and guarding single eggs on precarious ledges.

The sea around the boat was full of guillemots trying their hardest to make their escape, fizzing arrows of activity as they pushed forward through the sea. The smell was there too – warm, slightly acrid, and strong enough to make all the kids on the boat wrinkle their noses. I didn't mind it – I've almost grown to like it. Like a farmyard on a hot day, it's not the most pleasant of smells, but it is natural, and there is comfort in that.

At one point, the Pinnacles come away from Staple Island; massive sea stacks standing like a row of rotten teeth in the sea. The grey sky was visible through the gaps, and the white guano was patterned by streaks of green algae. Kittiwakes had taken up residence in the crevices above the tideline, but not at the top. On the tideline, cormorants were king, perching on the rocks closest to the sea and looking not unlike pterodactyls waiting for the next Jurassic Age, or a tasty fish, whichever came first. Cormorants have to dry their wings out after each fishing trip, hence all the hanging around. On the top, guillemots had colonised the upper town too but, as we moved around the Pinnacles, they were replaced by puffins.

The first sight of puffins drew *oohs* and *aahs* from the boat, like even the worst firework shows in November. The kids had forgotten the smell and the feigned seasickness, and the adults were all on their feet with cameras and phones pointed aloft. I love the way that puffins can do this; the mere sight of a puffin can raise a smile on the bleakest of days. It's their slightly comical look, their brightly coloured beaks and being so unique that causes this reaction, I think. It's no wonder that they are reproduced on tea towels, T-shirts and mugs, he says, writing this while wearing a puffin T-shirt and drinking out of a puffin mug that he's just dried with a puffin tea towel. Puffins make people smile. They may not be the UK's favourite bird – a recent national vote put the robin in the top spot, while the puffin came in at a harsh tenth, behind even the blackbird and the blue tit – but if the public knew the truth about the territorial, bloodletting antics of the robin, they wouldn't be so keen. The PR of the Christmas card image has done wonders, but do robins have their own cartoon? No. Puffins do. Enough said. Even Chris O'Dowd does the voiceover for *Puffin Rock*, so it must be good. Puffins, especially when combined with an Irish accent, are a sure-fire way to make people smile.

Puffins lined the top of the rocks in rows. We were seeing them from below, which was unusual; I was used to looking down on them from clifftop locations. This view accentuated their white chests, making them look regal and proud. The colour of their beaks was the only colour around, and, especially against the sky, made them look like extras from a travelling fairground. I was reminded that the red part of the puffin's beak has grooves in it, which can roughly indicate the age of the bird. This is not an exact science though, with youngsters having a single groove and older puffins graduating to a third or even a fourth groove. There is also a difference in beak size; only a fully grown adult bird will have a full, thick bill.

There were a great number of puffins at the Pinnacles, I was pleased to see, and the National Trust reports a stable population of around 43,000 breeding pairs on the Farne Islands. They've increased the frequency of their surveys from every five years to annually, to keep a steady eye on them. Unlike the guillemots, the puffins were calm and surprisingly quiet, seemingly friendly to one another.

The boat trip continued with a brief visit to the slightly gaudy-looking lighthouse on Longstone. Its tower is painted pillar-box red with a giant white stripe, and its nearby buildings were seemingly painted from the same pot.

Since 1914, grey seals have been protected and their numbers on South Wamses have grown into a sizeable colony. The seals were hauled out onto a cluster of low rocks, all adults, as the young wouldn't be born until winter. They inquisitively watched the boat pass, but were too lazy in the midday warmth to come across and investigate further. They were plump and mottled with black-and-white spots, which reminded me of an overfed Dalmatian that belonged to a dear old lady I once knew. A canoeist dressed in exactly the same red as his craft and trying to get a closer view, showed just how big the seals were. A grey seal bull can measure up to three metres in length. The canoe looked tiny in comparison, and I couldn't help but think that a shove from one of those fella's flippers would see the canoeist straight back to Seahouses.

We left the seals lounging around as if they were on a cooler Costa del Sol and headed for Inner Farne, where we would have the chance to look around. For a small fee, of course.

After dodging the terns and losing some blood, I walked hurriedly past the ruined 'Fishehouse' to the information centre in the stone-walled courtyard. It contained the usual informative displays courtesy of the National Trust, a table of skulls of the island's animal population, and a camera placed in a puffin burrow, which was currently unoccupied. There were also the usual schemes to persuade puffin lovers to part with yet more cash, but I was wise to such tricks and left the white chocolate 'puffin poop' well alone.

Opposite the centre was St Cuthbert's Chapel, requiring us to once again run the gauntlet through Arctic tern territory. The island's army of volunteers had marked every Arctic tern scrape with a pebble painted blue, each bearing a black number. I managed, despite putting my scalp at severe risk of another wound, and without disturbing the bird, to get a photo of a pristine Arctic tern sitting on nest number 100. I later tweeted this photo on social media, only to be asked how I'd managed to get the bird to pose for the photo. Social media sometimes brings out the strangest of people.

St Cuthbert's chapel has fared well, despite having endured the most inclement weather for centuries. Its current excellent condition is mainly due to the continuing care of the National Trust. The striking stained-glass window dates back to 1844. If you look closely, the window contains depictions of seabirds, but I couldn't spot a puffin. Saint Cuthbert himself lived on Inner Farne as a hermit and became so fond of his seabird neighbours that he had the foresight to put restrictions in place to protect them. Locally, eider ducks are still known as 'cuddy ducks' in his honour.

From the tower – atop which a fire used to be lit as a basic lighthouse – a path runs in a rough rectangle around the island. I followed it past the garden area, where a black-headed gull had made an unlikely and untidy nest on top of a lichened drystone garden wall. It stood proud on its flat messy sprawl of twigs, next to two spotty, fluffy chicks that were only a few days old. A short way down the wall, a family of eider, sorry, 'cuddy' ducks had made their home too. Mother duck was asleep with her head tucked under one wing, a pose that I know many parents with youngsters can identify with. Nearby, a bundle of chocolate-brown chicks cheeped contentedly to one another, among a fierce patch of stinging nettles.

The centre of the island was surprisingly green. It was covered with a dense blanket of vegetation, no doubt fertilised by the birds and watered by the frequent rain. The result was a thick growth, through which birds, including plenty of puffins, could be seen. It was fairly unusual to see puffins like this; I was much more used to seeing them on the sea or on cliff edges, and not in the centre of an island on fairly flat ground and in dense foliage. Towards the south end of the island and around the edges, the vegetation clearly thinned out, but the overall impression was of a surprisingly fertile place.

Walking towards the Churn along the wooden-planked paths flanked by a low, blue-roped fence, I stopped for a little while. The vegetation was lower here, and I could see tens of puffin burrows. Puffins were coming and going all the time, often returning with beaks full of sand eels. And this was the problem. Lesser black-backed gulls had figured out that an easy meal awaited them. In fact, it came with its own delivery service. I waited and watched. A puffin flew in with its well-earned catch of five or six sand eels. The sand eels hung out either side of its beak, their silver brilliance reflecting the weak sunshine. Sand eels are not actually eels at all, but small, iridescent fish usually a few centimetres long. They travel in shoals and can live up to ten years if they're not caught by the variety of seabirds, including puffins, that depend on their oily goodness, or fished from the sea in huge numbers by teams of trawlers. A gull was watching the puffin too and immediately headed towards it. The two met in a flurry, just as the puffin landed in crash-managed style. The puffin was straight on his feet, head down and scurrying. He twisted and turned around tufts of white-flowered sea campion, before ducking neatly down a burrow, leaving the gull angry and frustrated. Puffin 1, Gull 0. This was kleptoparasitism in action – one species stealing food from another. I suspected that the puffins were not always quite so lucky.

The path veered towards the top of the cliffs towards a clutch of white buildings that included the lighthouse. There were plenty of puffins here, and guillemots too. They lined the clifftops, the occasional sea breeze ruffling their head feathers. There were five or six birds to every metre here, and the noise to match. Further around the path, another colony of puffins could be seen on a greenish-yellow patch of rocks on the south of the island. There were fifty or so posing for the cameras, sunning themselves in the pale sunshine and generally doing puffin things such as preening, dozing and cawing quietly to each other.

The short lighthouse was enclosed by a low, white-washed stone wall. The gardens beyond the wall had become a favourite haunt of the terns, which had moved in in their dozens. I couldn't blame them. It was an attractive building, resplendent in a fresh white coat of paint and with the Trinity House coat of arms displayed proudly on its side.

I headed on towards the stack, where the path led straight to the top of a sheer cliff. There was a wire fence, but I was essentially in a colony of shags. They were everywhere, in corners of the angular rocks, their big, messy nests made of anything they could find. Twigs were intertwined with rotting kelp, string and plastic bags to form circular nests that looked like the floor of an unruly teenager's bedroom. Peculiar-looking chicks peered awkwardly from the nests, at various stages of development. Shag chicks are chocolate brown but appear to have stuck-on, joke-shop faces. The parents are equally odd-looking, with massive black rubber flippers for feet and, when seen up close, a deep emerald sheen to their black coats. They have mad, starey eyes and a stuck-up cowlick crest on their heads. They are slightly smaller than their cormorant cousins, with long black beaks with a yellow base, and sinuous necks.

There were several human children and their parents up against the fence, looking keenly at the birds less than a metre away. The parents had probably not anticipated having to explain what two of the more amorous birds had started to do in front of their children, nor how this was not exactly how the shag had got its name. At least shags don't peck the back of your skull, I thought, tentatively checking my scalp for holes left by the attacking terns.

7

Southern Puffins

May

Annet, Isles of Scilly, England

The *Scillonian 3* pulled out of Penzance harbour with a
blast of her horns. I was joined by families heading to the
Isles of Scilly for their summer holidays. Flip-flops, rugger
tops and straw hats were very much in fashion. Suitcases
were separated on the quayside according to which of the
five main islands they were destined for: St Mary's, Tresco,
St Agnes, Bryher, or St Martin's. My plan was to take the
Scillonian the twenty-eight miles to St Mary's, where I hoped
to find a friendly boatman to take me out to Annet and the
last few puffins in this southern outpost.

The Isles of Scilly have a long history and connection
with seabirds. However, where once there were hundreds
of thousands of seabirds, so many that they were used as

currency on the isles in the thirteenth century, there are now thought to be a mere twenty thousand seabirds. The Isles of Scilly Wildlife Trust estimates that 31 per cent of breeding seabird pairs were lost between 1983 and 2015/16 alone, a decline that is partly down to the brown rat. Scillonians are fighting back, however, with the Seabird Recovery Project working wonders to eradicate rats on St Agnes and Gugh. They even have a rat-on-a-rat hotline for any rats spotted, and volunteers check incoming deliveries for unwelcome stowaways. They take biosecurity very seriously there. I just hoped that I would get to see a Scilly puffin.

The *Scillonian* is a strange vessel, flat bottomed and prone to making everyone on board feel sick, so she's often referred to as the *Sickollian*. The number of sick bags provided and the proliferation of yellow plastic mops and buckets only went to show how real this was and why she doesn't sail during the winter months. It was May, so I remained cautiously optimistic.

I chose the airline-style seats of the main cabin, which reclined nicely. There is a helicopter service out to Scilly, but it's expensive, so the only options are the *Scillonian* or the unpredictable flights from Land's End or Exeter. Plus, the *Scillonian*, for all its faults, felt like an adventure, a real trip, even if it did include snotty children climbing over you and bewildered dogs barking themselves to sleep. A widescreen TV advertised the best of the Isles of Scilly and contained scenes befitting a Mediterranean island. I looked out of the cabin window to see a grey, sad sea and clouds cut from the upholstery of old Ford cars. The TV showed scenes of swimming puffins though, which made my heart jump.

I didn't care if the weather wasn't on my side for a change; seeing some of the UK's most southerly puffins would be enough of a treat for me.

The *Scillonian* chugged on, sending spray up against her sides. I mainly stayed seated, as my sea legs are not great, and I'd seemingly sedated myself in an attempt to avoid seasickness. I did make it to the café though, where they made boiling coffee (presumably to cause maximum scalding to passengers trying to walk with it back to their seats) and Cornish pasties. I love a good Cornish pasty – I've survived a few days on pasties, cream teas and cider; what else do you need? – and couldn't pass up this opportunity to grab one. I made my way back to my seat, succeeded in not scalding myself with the coffee, and settled down to my Cornish pasty just as we slipped past the village of Mousehole. A neighbouring passenger smelt the delicious combination of beef, vegetables and onions and turned a funny shade of green. Maybe it was too much for her. For her sake, I decided to eat the pasty as quickly as I could.

Puffins, like other seabirds, are vulnerable to pollution, especially oil spills. This was particularly apparent in March 1967 when the SS *Torey Canyon* ran aground on the Seven Stones reef between Land's End and the Isles of Scilly, releasing into the sea 119,000 tonnes of crude oil destined for a refinery off Milford Haven. The spill spread far and fast. The impact was devastating; seabirds were first affected by the oil itself and then by the experimental caustic detergents deployed in an attempt to disperse it. The oil clogged wings and beaks, preventing flight and feeding. Seabirds trying to clean themselves ingested the oil and detergents with fatal consequences. Estimates place the number of seabirds killed at between fifteen and thirty thousand, although no one truly knows the extent of the devastation.

The larger *Amoco Cadiz* oil spill in 1978 caused further devastation on the coast of Brittany, hitting some of the same areas already affected by the 1967 spill. Twenty thousand deceased birds were recovered, but many more are thought to have perished. The RSPB estimate the loss of puffins alone to be 85% of the breeding population along the Brittany coast. The impact was acutely felt in Les Sept-Îles, a group of small islands off the Brittany coast where the small population of puffins is still trying to recover to this day. We need to look after the *macareux*, as our French friends call the Atlantic puffin.

The Isles of Scilly are further from the English mainland than Calais is from Dover across the English Channel, and the *Scillonian* was not a fast ride by any means. I occasionally swapped my comfy chair for one of the outdoor park-bench-style ones and a bracing gulp of sea air. From there, I spotted the odd Manx shearwater and gannet, but little else.

The isles started to come into view – first the tiny bits and bobs of the Eastern Isles, and then, in the distance, the impressive white-sand beaches of St Martin's Flats. The sea was calmer here, but it was still overcast, so I didn't get the full 'we are in the Med, but we are not really in the Med' experience. Which was a shame. I was only in a T-shirt though, which was something I'd never managed to get away with when visiting Scottish or Welsh islands.

I thought about buying a straw boater, and maybe even some deck shoes.

There was anticipation in the air as we docked in Hugh Town, St Marys. The new quay was bustling with arriving passengers and others trying to continue their journeys to different islands. It was a confusing affair and, although I'd planned to grab my ticket to Annet from there, I couldn't make head or tail of what was going on. The system seemed to work around a blackboard stating which boat was going where and when, which was marked up every day by the St Mary's Boatmen's Association. I couldn't see one for Annet, so I headed down the quay into Hugh Town. Town Beach looked inviting. Just a quick dip of the feet to cool off after the ferry?

Hugh Town is the capital of the Isles of Scilly and is squeezed between two white beaches before the bulk of St Mary's, including the Old Town and the only A-road in the isles, which runs in a short circle. It felt like a British town that time forgot. Not in a bad way. There were very few cars, it was peaceful and there was a strong sense of community. I couldn't tell whether people still left their doors unlocked around there, but I suspected they did. One house had a colourful box of sea urchins on its wall and an honesty pot sitting next to it. The pubs were all nautical-themed – Atlantic, Slip Inn, Mermaid – and I hoped to make use of one later in the day, sharing a bench outside with locals supping beer. The Isles of Scilly have managed to increase the level of tourism, especially during the summer months, without losing the magic of such a wonderful place. They are a well-kept secret.

At Mumford's, a newsagent in the sense that they sold newspapers and everything else you could think of, the guy behind the counter told me that the *Guiding Star* would be heading out to Annet at 2pm. I paid for a ticket up front. It was a tiny slip of a ticket from a roll, like I used to get from the cinema as a kid.

I walked slowly down to Porthcressa Beach. The sand here was beautifully white, and even the limpet shells looked scrubbed clean. Outside the tourist information centre – now full of the disgorged contents of the *Scillonian* – I found a picnic table and shared my lunch with an inquisitive song thrush. Song thrushes were frequent visitors to my garden as a kid, but I hardly ever see them anymore, so I was more than happy for him to dart around my feet and to and from the table itself. His spotty chest was silky smooth, his eyes bright and keen. Song thrush eggs, if you ever get to see them, are brilliant blue, like sweets from a sweetshop. The song thrush was joined by a rowdy gang of starlings – footballers, my gran used to call them, on account of their flashy tops and gang mentality – whose noise and bother only served to attract a herring gull that stomped around like a grumpy old man and ruined everyone's fun.

You can't get further south in the UK than the Isles of Scilly and, as a result, flowers grow here that are usually found in much more distant climes. I spotted plenty of tiny colourful fields. The islands also have many species of rare wild plants, which can only be found on Scilly and nowhere else in the UK. These include orange birdsfoot and the mysteriously named least adder's-tongue.

A walk on the pristine beach gave up a cuttlebone from a cuttlefish. White and smooth like a bar of soap, cuttlebones give the shape to the cuttlefish, which isn't actually a fish at all but a mollusc. Cockle and limpet shells crunched underfoot and sand hoppers inhabited the black line of bladderwrack. I was hoping to see a Scilly shrew in the seaweed too, but no luck. Also known as the lesser white-toothed shrew, or locally as the teke, it's the rarest shrew in the UK. They live among the dunes and eat insects in the seaweed on the shore. It's a rather cute fact that baby Scilly shrews fresh from the nest will hold onto each other's tails, the front one holding onto a parent, creating a train of shrews as they head out for the first time.

While walking, I was half looking for LEGO from the *Tokio Express*, which lost her load during a storm in 1997. Some 4,800,000 pieces of LEGO were released into the sea, including many marine-themed items such as octopus, flippers and scuba gear, and they're still washing up on Cornish shores to this day. It's the dragons that are the most sought after. Discarded plastic causes huge problems for seabirds, including puffins. I'd seen it being gathered with nesting material by the shags on Inner Farne, ready to harm both parents and chicks. I've seen the sad remains of seabirds strangled and trapped by plastic waste, including rubbish and discarded fishing gear. Studies on deceased puffins from the Isle of May by the UK Centre of Hydrology and Ecology found their stomachs contained numerous plastic nurdles. Plastic is slowly and painfully killing our seabirds.

I didn't find any LEGO, but I did find a beautiful painted top shell. Shiny as a jewel, it reminded me of both a helter-skelter and a barber's pole. Do barbers have poles these days, or do they just rely on Facebook? That was not the point. It was a beautiful shell, no longer required by the snail that used to call it home, and I smuggled it into my pocket. LEGO dragons – who needed them? A dunlin caught my attention near some rock pools and, although I was more than happy to stay, I had an appointment with the *Guiding Star*.

The quay was still busy, like a maritime bus station, and the *Guiding Star* was just making her way in. She's one of the oldest boats still in operation in St Mary's and was built in 1933 in Mevagissey, Cornwall. She looked a beauty, in light blue and yellow and with seats so varnished and clean you could see your reflection in them. I was welcomed aboard by the captain, Joe Badcock, who told me that the boat had been his father's and in the family for forty-two years. That was somehow reassuring. There was a small cabin up front but no other cover. The clouds were still lingering, so I opted for the seat closest to the cabin. There were no other passengers, so I passed the time by chatting with Joe. He told me that Tresco was his favourite island but that he enjoyed the variety of taking different routes. He had some good news too – puffins had been seen on the water at Annet that morning. My heart jumped again. Southern puffins!

A few more passengers climbed aboard and although the boat was still far from full, we were off. Joe took us out onto the calm sea and into the sound between Samson to the north and St Agnes and Gugh to the south. Annet is a couple of miles to the west of St Agnes. A little closer, and he pointed out landmarks off St Agnes. These had wonderful names such as the Barrel of Butter, Little Brow Smith, Tins Walbert and Three Brothers. Pronounced in Joe's fantastic Scillonian accent, they all sounded like places where pirates had buried treasures yet to be found. As he got some speed up, I was doused with plenty of sea spray, even though I was seated near the cabin. I didn't mind, to be honest; it was a hot and humid day and I quite welcomed the refreshment. He took us out into Smith Sound and headed to the north end of Annet first. It was here that he'd seen the puffins that morning, while I was still aboard the *Scillonian*.

Annet is just over half a mile long and from the air looks a little like a diving fish. The tail is the north end, if you are struggling. I could see Annet Head and Carn Irish as we got a little closer. Beyond Annet is Melledgan, and beyond that the Western Rocks, or the Dogs of Scilly. There's one further outcrop at Bishop Rock, but then the Isles of Scilly have all but run out.

Joe put his local knowledge to good use, aware that puffins mainly nest at the north-east point of Annet, around Butterman's Point. I saw a few birds on the water some distance away, but I couldn't fully make out what they were. They looked suspiciously puffin-like in silhouette. Grey seals were lounging leisurely on the rocks at Minmow. They watched us glide past, fully hauled out on to the rocks.

And then I spotted him. He was just a few metres away, initially hidden by the dark blue swell. He was a perfect little specimen and was carrying six or seven fish in his brilliantly coloured bill. His chest was as white as the sea's own surf, and he stared intently, just as intently as we were staring at him. Male

95

puffins are slightly bigger than females – by up to nine per cent – but it's not really possible to tell them apart by eye. The kids on the boat, three of them, went crazy at the sight of their first puffin. The girls were *aww*-ing, while the boy was pointing and shouting to an unbothered dad. I loved seeing their excitement, and I shared it fully, but it is the fact that a lone puffin in this era of smartphones, apps and Minions could still whip up such pleasure in young people that gives me hope for the future. I was sure that this moment would stay with them for a long time to come.

In the Breton language, the puffin is known as *poc'han* and it's entirely possible that this is where they got their name, rather than from the description that their portly bodies are 'puffed' up. This puffed-up puffin turned slightly so that he was side on, and the sun caught the silver of the fish like a jeweller's window. Then he had had enough and he pushed forward with effort, before running across the surface of the water like a gymnast and then taking flight. I watched him go. He was heading for Annet alright, and his cargo of sand eels meant that there must be a puffling waiting for him. Another new generation coming through.

No one can land on Annet, or Little Agnes, as it was once known. It's protected, and with good reason. It was once home to the archipelago's main colony of puffins, with over 100,000 birds. Sadly, hunting, pollution, rats and a depleted food supply have meant that, according to the Isles of Scilly Wildlife Trust, the entire puffin population of the Isles of Scilly is, in 2020, estimated to be around 170 pairs. A lonely shag stood guard on a sharp stack of granite, seemingly in mourning in his sombre dress.

Joe kept searching, turning the boat this way and that, monitoring the radio and making calls to other skippers out and about. We found more puffins towards the south of the island, but only in groups of twos and threes. These puffins seemed more skittish somehow, less relaxed than the first one. They too had mouthfuls of fish but were quick to move on. I noticed that they were flying back to a line of exposed soil beneath a layer of pink sea thrift, which was patrolled by lesser black-backed gulls. Nowhere on Annet is particularly high, so you couldn't call it a cliff face really, but the puffins had made their home here. Below them there were a few razorbills. At night, Annet would also become home to small numbers of storm petrels – the only place in England where they breed – and Manx shearwaters, although they too are sadly reduced in number.

Rather than return to St Marys, as I presumed would happen, we chugged slowly over to St Agnes. Joe was full of apologies that we hadn't seen more puffins, but I was thrilled with those we had spotted. He moored us casually alongside the pier at Porth Conger on St Agnes. 'We've enough time for a pint,' he announced. It sounded like a plan, and the sun had come out too.

Disembarking here was like arriving at a tropical island. The sand was white and smooth, and the path up the hill was lined with wildflowers. The sea had turned a translucent aquamarine and was perfectly calm. It may not have had the warmth of more tropical oceans, but it had the look down to a fine art. It was difficult to believe that this was England. I wasn't alone in this – rare birds often get blown off course and find themselves unwittingly landed here.

Bees had returned from shelter and slumber and were busying themselves. A salty sea dog leant over the side of a boat to greet me, tongue out and tail wagging. The Turks Head is the most south-westerly pub in the United Kingdom. It had a rusty, squeaking sign hanging above the door, as all pubs should, and a life buoy nailed to the wall, either to remind us of its nautical setting or perhaps to lend to customers when they've over-indulged. It wasn't clear which. I stepped inside to a good old-fashioned British pub. It seemed rude not to have a pint at the bar. I raised my glass of Cornish Rattler to the last few puffins of the Isles of Scilly.

8

Puffins and Mermaids

June

Treshnish Isles, Inner Hebrides, Scotland

'Don't forget your sun cream,' said the B&B owner serving me a full Scottish breakfast, 'Or you'll have a wee pink facie.' He was probably right; he looked like a man who would know. He also clearly knew his way around a frying pan. Any more of that and I wouldn't be going anywhere; I'd end up sitting there feeling the breakfast congeal in my arteries.

The Isle of Mull was bathed in uncharacteristic sunshine as I made my way to Ulva Ferry, passing 'Please Slow Down, Otters Crossing' signs, although sadly there were no actual signs of otters on the shore. I'd have loved to have seen one, but they were evidently keeping themselves cool. Lonely oystercatchers roamed the shoreline, pipping their way around. Another sign in the green lanes read 'Slow

Down for Cats, Dogs, Children and Doddery OAP', which made me laugh. I'm not sure that laughter-inducing road signs are a good thing for road safety though. Highland cows, or should I say 'coos', barely lifted their giant shaggy heads as I passed.

I was meeting the *Hoy Lass* boat at Ulva Ferry, the place from which the ferry to Ulva departs, which turned out to be little more than a portacabin toilet, a few store buildings and a line of cars. There was a mongrel of a dog wearing his very own life jacket though, which caused me to smile. The sea loch here was beautifully calm, with barely a ripple on the surface. A wooden pier reached out, and the blue sky was leaking into the water. The *Hoy Lass* was waiting to take me to Lunga and Staffa. She'd had a handsome red-paint job, and on her lower deck there was a classroom-styled space with rows of chairs and slightly disturbing bright yellow walls. I was pleased to see that you could buy a mug bearing the legend 'Puffin Therapy'. Each mug entitles the new owner to unlimited scalding hot coffee from next to the captain's chair. I bought one immediately, of course. I later found out that Turus Mara, the company offering the tour, had had their puffin therapy officially – or thereabouts – endorsed by an expert. In 2011, Dr Nick Baylis, a psychologist and well-being expert, lent his opinion: 'Communing with the wild and with birds like puffins is as important as sunshine, or sleep, or Vitamin C. Puffin therapy is a great way to get that fix.' You can't argue with that.

We chugged slowly out into the loch, passing tiny islands topped by their own wonderful tartan of pink sea thrift and bluebells. Common seals watched us slide past, unconcerned by our presence. This was Loch Tuath or the Northern Loch. To the north of us, Mull continued on, its tranquil presence only disturbed by the hundreds of elderly day-trippers on coaches munching warm sandwiches and completing their knitting. To the south was the Isle of Ulva, with a well-wooded shoreline and Beinn Chreagach dominating the skyline. Ulva – from the Old Norse for wolf – sounded an intriguing little place and looked exactly wild enough to imagine that wolves were watching us from the dense covering of vegetation.

Just before the boat got to the island of Gometra, separated from Ulva by a narrow, thrashing chasm of sea, the captain cut the engine. It was extremely quiet. The waves lapped at the hull of the boat and someone poured themselves another coffee. The captain had spotted a white-tailed sea eagle on her nest. He pointed her out, sitting on top of a huge, messy nest the size of a dustbin, lodged in the bough of a long-deceased tree. The canopy was turning green around her but, for now, she was left partly exposed. She was peeking over the top.

White-tailed eagles were reintroduced to Mull some forty years ago, after being completely wiped out through persecution, and have done so well that they've spread to Ulva. The nests are used year after year by the same pair. I could just about make out her head in the massive nest, her feathers blending perfectly with the browns and greens of the bark and sticks in which she rested. She looked a bit grumpy, to be honest; perhaps it was time for the husband to take over; both parents incubate the egg, and he was not far away; we saw him in another dead tree some five hundred metres distant.

Both birds were absolutely huge. White-tailed eagles have the largest wingspan of any eagle, up to two and a half metres. I've seen white-tailed eagles before, in Iceland, and their size always amazes me. I could see the male bird's wide, almost white coloured head, on top of the even wider shoulders, and the dark-grey-and-brown feathers. He looked like a thick-necked nightclub doorman, even if the nearest club was some miles away. His thick legs and feet were bright yellow, like some kind of warning signal. He was as still as could be, staring out into the loch from his high perch. As he didn't take flight, I didn't get to see those majestic wings spread out, or the band of white on the tail that gives them their name.

We pushed on, with the *Hoy Lass* gliding gently past Gometra and out into the open sea. Things got a little choppier there, but she took it in her stride. We were heading for Lunga, one of the Treshnish Isles some sixteen miles from Mull. The Treshnish Isles are a tiny, isolated archipelago that almost no one has heard of. Staffa isn't strictly in the Treshnish Isles, but lies to the south. There was plenty of birdlife; black guillemots scattered from the boat, their plumage like shiny coal, and cormorants stood on broken jumbles of rock, sunning their wings to dry them before their next dive.

Landing on Lunga was not a simple matter. The captain had to perform some interesting manoeuvres to collect a floating pontoon, which he then gently placed against the football-sized boulders that made up the shore.

The boulders were slippery and it took patience to cross them without getting either a wet foot, or, worse, a broken ankle. As I made my way over them, it began to get easier on the rocks that were out of the tide's reach and, therefore, less slippery. Between them lay broken branches, desiccated seaweed and the bright green tufts of a long-forgotten rope; the remains of storms long since passed. I scrabbled on, showing my characteristic lack of balance to full effect, and was hugely relieved when I hit the coarse grass. As a welcome, Lunga had also provided a wonderful display of sherbet-lemon primroses, which stood out from the dark grass like autumn fireworks.

I made my way up a meandering path that led to the clifftops. The captain's words rang loudly in my ears: 'Don't stop at the first lot of puffins you see. I say that every day, but no one listens.' I was some thirty metres up and, as I rounded the final corner, I saw a wide, flat, grassy path. To the left, the cliffs continued upwards slightly but, to the right, was a view like no other. Hundreds of puffins lined the cliff edge, beyond which was a beautiful, cerulean sea. There were burrows right up to the cliff edge, and up to a couple of metres inland. The soil here was soft and yielding and clearly perfect for puffin architecture.

I did exactly what the captain had said not to – I stopped in my tracks. I couldn't quite believe my eyes, especially after the lack of puffins in some of the other areas I'd visited. I dropped to my knees and stared out towards the bustling line of puffins. They were not at all alarmed by my being there, and I was extremely careful not to frighten them or crush any burrows beneath me. I was within a few metres of the nearest bird, who was merely sunning himself. He turned this way and that to check himself and to see whether there was any danger, but seemed calm and relaxed. The sunlight glinted through his bill, which glowed orangey red. I realised I was smiling so much that my face was beginning to hurt.

I lay down on the lush grass and just watched the puffins in front of me. They were completely unconcerned by my presence and, as in Skomer, I wondered if they were using the fact that humans were there as an effective tool against predators. The opposite could also so easily have been the case; there are examples of puffin colonies being disturbed by tourism and repeated viewing.

The puffins were extremely busy. Some were nest building, pulling strands of grass and dried plants before scuttling down burrows to line their bedrooms. Others were billing. There are a lot of similarities to kissing, it has to be said, even if it is closer to 'Eskimo kissing', as we called it in our playground – rubbing noses together rather than mouths. Try it.

The billing couples in front of me were getting a fair amount of attention from other, loafing puffins, who stood around to watch. A peculiar display was taking place where three puffins were billing, and I struggled to untangle exactly what that meant; I can't say I've experienced the human equivalent. It was attracting quite a bit of interest, and not just from other puffins. Squabbles and disagreements can occur between puffins, often in relation to finding a mate or ownership of a burrow. This can escalate to a show of dominance, with puffins growling, locking beaks, biting, and with wings outstretched, seemingly trying to push each other into the ground. It rarely ends in death but, rather, with the least fortunate puffin trying to make good their escape with their pride intact.

From my angle, I couldn't stop looking at the puffins' feet and legs. Firstly, they were a beautiful Fruit Pastille orange, and, secondly, the puffins often appeared to be dancing, lifting one foot up before the other.

I'd forgotten how noisy puffins are, and how their low, peculiar caws and moans are so unique – the last sound you'd expect to come from a puffin. Imagine a summer beer garden at an English pub. Your mate has been drinking all afternoon and has popped to the bar for another couple of pints of Twisted Spire ale and some Walkers crisps. He arrives back at the picnic table you're sharing and places the pints on the table. He goes to sit down but misjudges the distance and ends up sitting in a surprised heap on the warm grass. He lets out a low, peculiar moan, which you hear between your own snorts of laughter. That noise he let out? It's exactly the same as a puffin call.

There were a couple of razorbills in the mix. Slightly bigger than the puffin, they are also more boisterous. This pair made a handsome couple, complete with white racing stripes running down and across their thick beaks. Razorbills are monochrome until they open their mouths to reveal a shock of bright orange. This couple were also amorous, and spared no blushes on getting down to business right in front of me. It was probably time to move on. The captain was right, and my fellow passengers had started to catch up with me. I also wanted to find somewhere to have some lunch.

I managed to find a quiet spot to grab a bite to eat. It was next to a small rockfall, which a couple of puffins were standing guard over. There was clearly a bumblebee nest nearby; bees kept visiting and I could hear them busily buzzing away. I'd brought some food with me and I couldn't wait to get stuck in before the sandwiches went warm and soggy. Is there anything worse?

Back on the boat, and we were on to the next place, as much as I could have spent the entire summer on Lunga alone. Sitting south-east of Lunga, Staffa appeared side on, with its unusual layer of basalt columns looking like the filling in an overstuffed sandwich. The columns were broken towards the bottom, leaving their flat, six-sided pieces bare, like spilt coins. Staffa looked stunning in the bright sunshine with its unusual reflection repeated in the endlessly blue sea.

We were off to Fingal's Cave first. The boat edged bow first into the cave, which was made up of rows and rows of the impressive basalt columns of ever-increasing heights. Other columns were seemingly hanging from the roof, reflected in the translucent seawater beneath us. Known as An Uaimh Bhinn – the Melodious Cave – in Gaelic, it's well known for its natural acoustics, most famously for inspiring Mendelssohn's *Hebridean Overture*, a point the captain tried to press home by blaring the first few bars through the ship's crackling, elderly tannoy system. It was fairly impressive as natural sea caves go, but I was keen to go ashore and get exploring.

We landed at the jetty at Clamshell Cave, and I alighted as quickly as I could. I was met by a stone structure marked 'Donations', which I thought a bit presumptuous, but I kept it in mind for later. There was a well-worn path from the jetty down to Fingal's Cave on the island's south coast and I decided to take this route first. The National Trust for Scotland had kindly provided a handrail and even nonslip paint in some of the more treacherous places, which I appreciated as I made my way to the cave, using the hexagonal blocks as stepping stones. These really were amazing to look at, and I wondered how cooling lava formed such perfectly symmetrical shapes over and over. The black steps sloped away, slowly dropping down into the calm blue sea. There were a couple of tiny isles near to the shore, where the columns resurfaced again, but these were more tightly packed and at different angles, like children's Play-Doh squeezed through a mould and then pushed to one side. Others reminded me of plastic hedgehogs in kids' party bags.

As I rounded the corner to enter the cave, I saw something so surprising that it caused me to stop dead and my heart to leap into my mouth. A mermaid. Sunbathing on the rocks in front of me. Her head was back, she had sunglasses on and light brown gently curled hair. Her soft, white skin was exposed except for a bikini-type top, and she was perfectly still. I moved slowly towards her, not wanting to startle her but, at the same time, trying not to appear as if I was creeping up on her.

One of my boots scuffed a stone awkwardly and her eyes opened. She turned to face me. 'Hi!' she said. She was really pretty, but I thought it odd that she spoke English with an East European accent. This wasn't how I imagined a mermaid would speak. She was watching me now, so I regathered my composure and continued walking towards her. As I did, her bottom half came into view, revealing shapely legs but, disappointingly, no scaled tail. I'm sorry to report that this wasn't a mermaid but a sunbathing kayaker whose kayak and rumpled wetsuit were lying next to her.

'Hey,' I said, as I moved past her and into the cave. I'd never been so crestfallen to see a partly clothed female before. I'd been sure that this was a siren of the sea, not someone eating a sausage roll in the sun.

After having a good look around the stunning cave and repeatedly shouting 'hello' at myself, I started making my way back, trying not to disturb the mermaid. I could see the island of Iona in the distance, famous for its centuries-old abbey, but I had to keep an eye on my feet for fear of twisting an ankle.

Back near the boat, I climbed up two steep sets of steps to reach the flat top of Staffa. I was surprised at the length of the grass, which clutched at my ankles as I walked. The delicate yellow flowers of tormentil poked through just off the path. The only thing in the blue sky, other than the blazing sun, was a contrail drawn by an aeroplane. I felt my nose burning and pressed on. I'd been told the puffins were on a slope on the north end of the island, past Goat Cave, so I followed the path running in that direction. The terrain was lightly undulating, but the blue sea was never out of view. It was a nice enough walk, and solitary, as my fellow sailors had either chosen to remain on the boat or were lounging around on the grass near the top of the steps.

A pink buoy on a fencepost in the distance denoted where the path passed perilously close to the edge, just above Goat Cave. The not fully leafed brambles reached out to grab at my boots, and I really should have been watching my step. I wasn't though. I'd spotted a wonderful thing. In front of me was a large, gentle slope that rolled into the pearlescent, Mediterranean-like sea. It was covered in beautifully lush, thick green grass a few centimetres high. Hiding in the grass were tens of puffins. They were entirely unperturbed by me and allowed me to sit nearby. They were a busy colony, frequently to-ing and fro-ing from the sea below. I loved the way they came in to land, with wings outstretched and slightly arched, feathers on each end of the wing pointing out, feet splayed as air brakes and their powerful white chests fully on show. One ruined the majesty by firing out some poop as he came in, but no one seemed to mind.

I couldn't get my head around how well the birds seemed to be doing there. They were clearly thriving. I started to amuse myself by playing Puffin Connect 4. The rules were simple: in the comings and goings of puffins you had to get a row of four together. It was a great little game, but one that can now be played in fewer and fewer places in the UK.

The puffins had a wonderful shimmer to their black backs – when it caught the sun, it was just magnificent. One bird pointed his beak to the sun as if to admire the clear blue sky or just catch a few more rays. Puffins returned with beaks full of thin, translucent sand eels, indicating that there were chicks beneath the grassy surface, hidden away in damp burrows. Puffins eat off a menu of various species of small fish but mainly favour sand eel, sprat and other juvenile fish such as herring and hake; larger fish such as the butterfish can be too big for chicks to swallow.

Puffins have some clever tricks to avoid making multiple return journeys from burrow to sea when fishing. Their upper beaks are specially adapted with rows of tiny reversed spines, called denticles, to hold multiple fish in place; further adaptions allow the beak to close in parallel like a clamp, unlike our own hinged version, and elastic-like sides so that it can open much wider. This lets puffins hold on to their slippery cargo, both underwater while catching yet more fish and back on land when delivering the catch to waiting chicks. The Staffa puffins were proving very successful at fishing. Trips to the sea were frequent and they returned with eight or nine fish at a time.

Puffins are expert swimmers, often appearing much more at home in water than in the air. The sea there was so clear and still that I could see the puffins darting around just beneath the surface, arrows of electric bubbles off on another mission. In fact, I could see all the way to the sea floor; the water glowed where the white sand reflected the sun, and was swimming-pool blue in other places. Puffin dives into the ocean to find food can last up to two minutes and at depths of up to sixty metres, although puffins tend to prefer several shorter, shallower trips. I could see the puffins using their powerful wings to cut through the water, just like the very best athletes.

There were tufts of sea pinks on the cliff edge, but mainly it was grass. The puffins used this to their advantage, peering through the blades at me with curious and inquisitive glances. It seemed like a fair swap. One clump of thrift stood proud of the cliff and was round like a Christmas bauble. A puffin was standing on top of it, almost like a sentry, twisting his neck this way and that to get the maximum view. Beneath him, the tide had revealed a series of skerries and luminous green-tinged rock pools where an oystercatcher was on the hunt for lunch. Another puffin was busy gathering grass in his beak, making it clear that a puffin's work is never done.

Sadly, and slowly, I made my way back to the boat and grabbed a seat at the back on the lower deck. I planned to sleep on the trip back to Mull, lulled by the gentle motion of the waves. I could feel my face burning already. The guy at the B&B was absolutely right, I did have a wee pink facie. We were just waiting for one last passenger. The boat rocked as she stepped on board. I recognised her. To my amazement, it was the mermaid. She sat nearby and smiled at me. 'I'm just catching a lift back,' she said by way of explanation. I nodded off gently, dreaming of mermaids and stealthy puffins watching me through lush green grass.

9

Holding Patterns

June

Isle of May, Fife, Scotland

James handed me a full set of waterproofs that were mainly banana yellow in colour and said I should put them on. I hadn't expected this, to be honest, and when everyone else on the quay was wearing shorts and flip-flops, it felt slightly odd. James, a young Scottish guy who had the same soft accent as Ewan McGregor, was our guide for the day. I was taking the RIB (rigid inflatable boat) from the Scottish Seabird Centre in North Berwick. North Berwick is a delightful seaside town not far from Edinburgh and was recently named the most expensive place in the UK to own a home by the sea. I could see why; it looked to be a fine place to live. On my way there, I'd passed the ruined Tantallon Castle, which was completely surrounded by bright yellow fields of rapeseed.

The Seabird Centre didn't open until 10am, which was a real shame, as I was dying for a bacon sandwich and a mug of hot tea. It seemed pleasant enough though, especially the sculptures outside. These included a seal perched on the rocks next to the sea and a giant if slightly scary Arctic tern outside the timber-clad building. I was most taken by 'The Watcher' by Kenny Hunter, a full-size sculpture of a man in waterproof clothing and wellington boots looking out to Bass Rock through a pair of binoculars. It was very lifelike, even down to the creases in his clothing. I very nearly bid him a good morning.

Once fully togged up in yellow Gortex and red life vest and fully briefed on all the health and safety guidelines, we were led down to the boat. It too was bright red, with twelve saddle-style seats, and a small cabin at the back, into which James disappeared after telling us that the swell was probably worse than it looked – the harbour area was particularly calm – and that we should think of the saddle as being like that of a horse. It was not the most calming thing I'd heard.

The Isle of May is some nine miles out into the Firth of Forth. I was fortunate that the sun was shining and the storms of the past few days had long gone. Even so, once the RIB picked up speed and hit the open ocean, it didn't feel at all calm. It felt like a rollercoaster, with the gentle ups followed by crashing downs. This wasn't like riding a horse, more like a bucking bronco. The yellow waterproofs came into their own, although they didn't protect my face from the frequent splashing of cold, salty seawater. I could taste the salt on my lips. It was certainly invigorating, if nothing else.

We were joined on our journey by several gannets from nearby Bass Rock. Curiously, they seemed to be flying alongside the boat rather than away from it. They were busy birds, carrying seaweed and other debris back to their nests. James promised us that we would visit Bass Rock on the way back.

As we got closer to the Isle of May, the gannets started to disappear, and were replaced by puffins, guillemots and razorbills. The isle proved to be bigger than I'd expected. Once around the point of South Ness, the boat idled for a short while, while the captain considered his next move. We were at Kirkhaven, a narrow channel that runs through treacherous rocks between an area known as Ardcarron and the Pillow. The Pillow is the polar opposite to how it sounds. It's all spikey, sharp black rocks pointing out of the boiling sea, which was currently an angry light blue, like a malevolent bubble bath. The rocks were covered in a cream foam produced by the raging sea. The captain made his move and, with speed and tenacity, he hurtled us through Kirkhaven, delivering us safely but breathlessly at a small jetty. It was exhilarating.

There was a brand-new visitor centre at Kirkhaven, which we used for stripping off our yellow outfits before quickly skipping outside like a gaggle of excited schoolkids. The visitor centre had one of those trendy new roofs that supports its own environment, only this one was special, designed to replicate the conditions preferred by terns for breeding. It was obviously working; the mixture of gravel and grasses had so far enticed nine pairs of terns to nest directly on top of the roof.

A Scottish Natural Heritage ranger would normally meet any new arrivals, but he was with another boat in Altarstanes on the opposite side of the island, so it was left to James to brief us on the dos and don'ts. The first of these – as usual – was to always stay on the path. This wasn't all that surprising, but to demonstrate why, James pointed to his feet. Only a few centimetres away from the path in front of him, in a low patch of nettles and dandelions, was an eider duck. She was incubating her eggs on her nest, and if James hadn't pointed her out, I doubt I would have noticed her, so perfect and effective was her camouflage. Around her was a circle of grey, cloudlike fluff – eider down. It is this stuff that they make duvets from, mainly from eider populations in Iceland.

James gave us a choice – we could either remain with him for a short tour of the island or make our own way around. He promised sincerely that, either way, we'd get a lift back on the boat. I opted to stay with him. He was an amenable chap and was clearly knowledgeable. He told us, for example, that the bamboo canes around each Arctic tern nest in the old walled garden area were an anti-gull device to prevent the theft of eggs, although the gulls would eventually figure this out and try an assault from the ground. Rangers had a plan to foil this too, in the form of a low-hanging line of bunting, not for celebration but protection.

As we rounded a corner on the Burrian, I saw my first Isle of May puffin. She was perched on a tall granite block next to the path. The group stopped to observe her for a short while. She was a perfect specimen, and her welcome to the Isle of May was loud and proud.

Puffins are not all the same size. There are differences between countries and even colonies. Puffins that breed in colder locations such as Greenland, Iceland and Norway tend to have larger bodies than their southern counterparts, which makes sense when you consider the temperatures they have to tolerate. However, it gets stranger than that. A study by MP Harris in 1979 found a significant difference between the sizes of puffins from St Kilda and the Isle of May. Puffins from St Kilda were smaller and lighter than those from the Isle of May. This was attributed to environmental factors, but Harris noted the peculiarity that a puffin from one side of Scotland could be larger than a puffin from the other.

After a few minutes, our puffin greeter left her perch and headed for the sea. We moved further down the path to where she had been. The gentle grassy slope to the left had been determinedly mined by puffins. It was riddled with burrows. Slowly, puffins emerged and departed for the sea. It was evident that incubation was going on underground. The female puffin lays a single white egg, which she and her partner take turns to incubate, doing alternate shifts for up to forty-five days. The off-duty puffin spends that time fishing out at sea. Using a wing, incubating puffins press the egg against special brood patches on their undersides; these keep the egg toasty warm until hatching. It was mind-boggling to think that every one of the myriad holes in front of me was likely filled with an expectant parent on a single egg.

Further down Holyman's Road, we continued beneath a delightful white-painted stone-wall archway that carried

McLeod's Path atop it and framed the blue sky perfectly. The island was resplendent with white sea campion and the intense yellow of celandine. I stopped to take a photograph. Stepping through the archway felt like stepping into the photograph itself, especially as beyond it was Low Light, a converted lighthouse and cottages. A washing line swung in the breeze and, beyond that, the lighthouse tower could be seen. Some of the buildings are used as a bird observatory, but you can stay there too. Since the refurbishment in 2014, it even has a flushing toilet. I was sorely tempted to spend a week there. Scratch that. Six months.

We followed McLeod's Path up over the arch, away from Low Light. Here, the path had whitewashed brick walls on both sides, an essential aid during the island's many fogs or on dark and stormy nights. After a brief intersection with High Road – again, not an actual road – we arrived at the impressively gothic Main Light, a lighthouse with serious attitude, looking more like a fairy-tale castle than a lighthouse. Clearly being refurbished, a decorator had left the door slightly open and I could see a marble fireplace and ornate staircases inside.

The main spot for puffins on the Isle of May is known as Bishop's Cove. Approached from a path that led off High Road, it was easily a highlight. The path ran alongside three small, boggy ponds – Three Tarn Nick – where several families of eiders had made their home. Eider mums have seven different calls, James told us – but most of the group was too distracted to listen. One of the eider ducks was taking her two tiny balls of eider fluff for a walk. They were impossibly cute, though clearly vulnerable – only a snack for a passing great black-backed gull. That's why eiders have so many chicks, because very few actually survive. Call it a strategy. They also create a crèche of chicks from different mothers and move them en masse, under the safe gaze of several mother ducks.

Beyond the tarns, we passed through a beautiful shallow valley covered entirely with flowering sea campion. I was as excited as a schoolchild seeing the first settling snow. That might have just been the anticipation though. I walked slowly down the valley, admiring the flora as I went. I could hear puffins and, if I strained my eyes, I could see them flying too. I quickened my pace. I could feel that I was already smiling.

The valley ended in a number of natural stone steps that led to a plateau. This was Bishop's Cove, some fifty metres above the sea. It was absolutely full of seabirds. In front of me, I saw several puffins and guillemots. Guillemots, James told us, take their name from the French for William. I loved the idea that they were all called William. But, again, no one was listening. There were puffins.

One was peeking at me from behind a barrier of rocks. I was reminded instantly of the cartoon *Puffin Rock* and I heard Chris O'Dowd's voice in my head. At the edge of the cliff, I looked down, expecting to see sea, but instead there was a small ledge inhabited by a pair of guillemots. The Williams were being as noisy as ever, but as they moved, I saw the light blue of an egg, which, thanks to the wonders of evolution, was shaped in such a way that ensured it wouldn't roll off this or any other such ledge. The parents were fiercely guarding it anyway, and I felt sure that this egg would be successful.

The main spectacle, however, was out to sea. There were puffins everywhere. They wheeled through the sky in formation, filling it with the orange beaks, pointed wings and flap, flap, flap of their flight. They flew right over me, almost past my face. Each time an individual circled past me, I could see them looking at me with intelligent eyes, weighing me up as a potential threat. I felt as though I could have reached out and touched one, and yet the other side of the circle as far out to sea. There was a definite pattern to it: out one way to collect food, and in the other to return to burrows. The circle was huge: as wide as one of the old gasworks you see in industrial towns across the UK and probably three times as high. It was dark with circling birds, circles on top of circles, and I could only think of the holding patterns of aircraft above Heathrow.

James and the group moved on, and I was sorely tempted to stay. Like a sheep, I followed them back to Main Light, from where we made our way down Palpitation Brae, a narrow, severely steep path that surely got its name from someone heading up it. I was glad it was downhill for me. Palpitation Brae ran between a low wall on one side and a thick, rusting pipe on the other. I was amazed to discover that this pipe used to be connected to the North and South fog horns and that a series of tanks at the bottom of the hill delivered compressed air to both.

James led us back to the visitor centre and we were let off the leash. We had an hour before the boat left, so we had time to explore on our own. James had some suggestions – South Ness, with the peculiarly named Lady's Bed, which was close to Willie's Hole, innuendo fans will be glad to know. The north end of the island, including Rona (an Old Norse name for seal) and North Ness, was out of bounds but would have been near impossible to get to anyway, thanks to a bridge collapse.

Besides, there was only one place I wanted to be – back at Bishop's Cove. I hotfooted it back up the paths to the cool, damp valley and then down the steps to Bishop's. It was deserted. I had the place to myself. Well, almost. There puffins and guillemots and I could see some razorbills that I hadn't noticed before lined up on the nearby cliff face. To my left – towards Slipped Disc (no, really) – I could see ten or so puffins on the very edge of the cliff, waiting to join the flying masses already in the sky. I took a seat on one of the low rocks near a clump of pink thrift. The sun was beating down on me and, for a moment, just for a moment, I was happy. There was no better place to be.

This feeling did not last. On the return RIB journey, the sky darkened and the sea became choppy. The RIB was tossed back and forth. James kept his promise and took us to Bass Rock, the largest gannetry in the world, home to 150,000 pairs. It was a white, seething mass of gannets. There wasn't a single gap unoccupied by gannets. Plenty of others have sung the praises of Bass Rock but, in that cold weather, and with the boat positioned just beneath the ruins of the island's medieval prison, complete with birds nesting in its old barred windows, it felt entirely apocalyptic. The gannets swirled and swooped. Close up, they're peculiar birds, with their beady, blue-rimmed eyes, yellow heads, huge white bodies and dagger-sharp beaks that look like they've been drawn on with a black marker pen. I did not enjoy their presence. It was cold and gusts of wind were blowing seawater down the top of my waterproofs and soaking my chest. I was shivering.

We returned to North Berwick. I handed my waterproofs back to James and thanked him kindly for his help. The sun had returned, and I got the feeling that Bass Rock was somehow cursed. To help with this I made an on-the-spot decision. I would seek redemption in a nearby church. It just so happened that the church was now a restaurant near to the Seabird Centre, and its congregation was there to eat lobster. I ordered half a local lobster with chips and rocket salad and a pint of cold local beer. I took a seat not far from 'The Watcher' gazing out to Bass Rock. My food arrived and I wolfed it down. It was delicious. Maybe I was religious after all.

10

Rhubarb and Puffin

June

Dunnet Head, Caithness, Scotland

The only good thing about John O'Groats was the slice of rhubarb-and-custard tray-bake from a café called Stacks, presumably after sea stacks and not the quantity of food on the plate. It was really delicious, a slice of Scottish summer on a plate. Rhubarb and custard is always better cold anyway. I was reminded that I once saw a Faroese recipe for roast puffin and rhubarb jam.

John O'Groats was truly awful. A smatter of low-slung buildings sat on the coast, scruffy and unkempt. The place was obsessed with being John O'Groats, the other end of the End to End from its southern counterpart in Land's End, Cornwall. Land's End pretends it is the most southern part of mainland Britain just as John O'Groats pretends to be the most northern. There's still a ferry to Orkney from there, and it was from a Dutch ferryman on that same route that the town took its name – Jan de Groot.

The buildings contained mostly different tourist souvenirs; one even disguised itself as a tourist information office but was full of the same gubbins. In another shop, there were plastic seabirds, all covered in dust and spiders' webs. The shopkeeper stood in the doorway, presumably to stop anyone entering and ruining her display by making a purchase. Real herring gulls patrolled the car park and scavenged for chips while children tried to chase them away.

Motorcyclists kept turning up and posing for photos next to the dilapidated signpost pointing out, from beneath multiple scruffy stickers, that it was 3,230 miles to New York. They laughed at each other and then moved off to a café, squeaking in leathers as they went. A Reliant Robin pulled up with a Land's End to John O'Groats sign on top. The driver looked like he'd done it all in one go.

I felt like a spoilsport, but it was such a tawdry little place. I really didn't like it. I finished up my rhubarb and custard and headed to Dunnet Head. That's the real northernmost point of mainland Britain, but don't tell the motorcyclists, End to Enders and holidaying families. They can stay in John O'Groats. They are welcome to it. Somewhere, Jan de Groot is slowly spinning in his grave as another tourist buys a cheap pen with his name misspelt on it.

The winding country roads led to Dunnet, a small village on the headland just south of St John's Loch, where fly-fishers stood thigh-deep in coal-black water, trying to entice Scotland's finest brown trout to take a nibble on artificial insects. I met Kate at the north end of Dunnet Beach, a sweeping bay of bleached white sand favoured by surfers from around the world. There was a caravan site there and a café.

Kate runs Caithness Wildlife Tours, which was exactly what it said on her smart silver people carrier, complete with a huge smiling puffin. The signs were good. I'd found Kate online, and she'd squeezed me onto one of her popular puffin tours at the last minute. It also meant that I got to ride in the front with her, giving me extra insight into the extraordinary Caithness countryside.

One of the two couples on the tour were on the North Coast 500, a tourist-office invention promoting a route around the Highlands, especially by motorbike, hence the shiny leathers they were wearing. Kate was a marine biologist from New Zealand with a kind face and black hair. She had an infectious laugh and a deliciously soft New Zealand accent – 'yez' rather than 'yes' –that reminded me instantly of the comedy duo Flight of the Conchords. As the comfortable but worn van made its way to the hamlet of Brough, she told me that she'd arrived in Scotland thirteen years ago with her husband and still hadn't really worked out a plan. Apart from finding puffins out on Dunnet Head, that is.

A brick bus stop had a spray-painted arrow towards Dunnet Head and a surprisingly good interpretation of the Ordnance Survey 'landmark' map symbol. Just outside Brough, the single track turned sharp right and a slipway took us through dense cow parsley to a small bay with an islet. This was Little Clett, and Kate had bought us here to see the local seal population that had hauled out. There were four grey seals and six common seals, all of whom were unperturbed by our presence, merely watching us from on top of their rocks. A pair of black guillemots flew in circles around the bay, while oystercatchers peeped to themselves on the rocky shore below. Kate gave us time to enjoy the view, telling us snippets about the seals, before gesturing for us to get back on board.

I could have stayed to watch the smart black guillemots for a while longer. They are intriguing little birds although they only wear their black outfits during the summer months, switching to white and grey in the winter. It's said that black guillemots can catch fish with the left or right side of their mouths, and that this allows you to tell if they are right- or left-'handed'. I'm not sure why you'd want to know this or what you would do with the information anyway.

I'm intrigued by the Scottish names for birds. The name for a black guillemot is tystie, which comes from Old Norse, and you'll probably know of bonxie for great skua and, of course, tammie norrie for puffin. There's often more than one word, or local variants. For example, other names for puffins include tammie cheekie, cootrie, coulterneb, buthaid, bougir, cockandy, ailsa parrot, bass cock, sea cockie, bottlenose, pope and lyre. There are plenty more too, mainly from Shetland and Orkney, all equally poetic and intriguing: alamootie for storm petrel, maali for fulmar. Maa means gull, so herring maa is obvious, and there's hoodie maa for black-headed gulls, and peerie (small, tiny) maa for common gulls. A razorbill is a sea craa, or sea crow, while a gannet is a solan or solan goose. Erland Cooper, a musician from Orkney, has released a record inspired by his homeland, each track being named after a seabird, and the composition sometimes featuring the birds themselves. This introduced me to shlalder for oystercatcher, cattie-face for an owl, and the wonderful moosie-haak (mouse hawk) for kestrel. It's a wonderful piece of music. Of course, these names vary wildly across Scotland and become interchangeable or fall out of use, but I think they are really rather lovely.

We took mainland UK's northernmost road – the incongruously named B8555. It led over the surprisingly flat marshland of the Moss of Dunnet, where there seemed to be another loch every few metres – Courtfall Loch, Black Loch, Sanders Loch, Long Loch, Loch Burifa, and then they clearly got bored of naming lochs, as the next group was simply called Many Lochs. The marshland was pretty with tiny clouds of bog cotton, which was growing in huge swathes, reminding me of the last snows of winter. Kate stopped to point out the bright purple heath orchids and marsh orchids, both of which were doing well in that northern outpost.

On arriving at Easter Head, Kate parked up just short of the car park and lighthouse. We decamped from the van, collected a selection of binoculars and telescopes and made our way down a drystone wall bleached almost white by the wind and rain. A skylark sang to us as we approached the clifftop. Beneath us, soft moss gave way to tormentil and sea pinks; on the clifftop itself, ragged robin added pretty splashes of pink, which, mixed with bog cotton, resembled a bag of spilt marshmallows. Rose root looked like an exotic.

At the top of the cliffs, the rock was showing bare, and I could smell and hear the seabirds before I saw them. There were razorbills and guillemots in abundance along the bottom of the sheer, craggy cliffs. The screeching guillemots could be heard distinctly over everything else. I miss that sound when visiting out of seabird season – abandoned cliffs that are eerily quiet. To me, guillemots are the sound of coastal spring and summer.

The day was overcast and cloudy, but I could still see the low grey cliffs of mainland Orkney, and even the Old Man of Hoy, as straight and tall as a water tower. What really drew my eye though was the number of birds in the sky, right over the Pentland Firth. There were tens of thousands of them; puffins flying in huge wheels had been joined by their auk cousins, and there were kittiwakes, fulmars and gulls too. They were busy, and the cliffs were bustling with life.

Kate showed me where to look for the puffins. As at Bempton, they didn't burrow at Dunnet Head, but made nests in the cracks and crevices just below the clifftop. She pointed out four beautiful specimens that seemed to be holding a meeting, which then had to be hurriedly terminated when a skua flew over, hearing their puffin plans and throwing the agenda out altogether. Another was having an afternoon nap, head tucked under wing, with a single eye peeking out for anything amiss. Kate said that breeding had been delayed by the unusually cold weather that spring; she'd been comparing the current numbers with photos from previous years. They seemed to be making up for it now though, as another popped his head out of a stony crevice that he had made his home.

The cliffs were alive with birdlife, stacks were chock full of birds, and each crevice and ledge provided another opportunity for a nest. The rest of my group were content to watch a gang of puffins in one spot, so I walked up the clifftop towards the wall. It was covered in mint-green lichen that was almost hair-like in places. I found a natural seat on the rock, an armchair with arms of tufted sea thrift. From there I could see along the cliffs and also had a view of a sleeping puffin. A stonechat bobbed on past me, stopping on a stile to announce its presence. There were wheatears too; they're similar in size, but once you know that their name has nothing to do with ears but is a derivative of 'white-arse', it's difficult to keep a straight face or forget this fact. Kate soon scooped us all up and we trudged back to the van. I hopped inside, and Kate told me where I should eat that night and where I definitely should not. She was full of useful information, and I enjoyed her company. I've never been to New Zealand, but I could see why she'd stuck around. My face glowed in the warmth of the van as we wound through lanes with boats and buoys on the verges.

Duncansby Head sits in the furthest north-east corner of the British mainland, where the Moray Firth meets the Pentland Firth. It's a wild, unforgiving place, much more so than Dunnet Head. The lighthouse here was fenced off and surrounded by menacing thistles. Duncansby Head was known as 'hell's mouth' to sailors, due to the fearsome tides and, even to me, it looked like the end of the world. An elderly couple were sitting in a Vauxhall Corsa, either too scared to get out or not strong enough to open the door against the wind.

From directly in front of the car park, a deep chasm dropped down to the sea. One wall was completely covered in sea pinks, like the sort of hanging garden you get in fancy hotels, and the other was surprisingly lacking in flora or fauna, with just the odd herring gull watching me warily. I guessed that most visitors didn't get much further than this, simply ticking the box, which was a shame. I walked down the path to the stacks. The grass was less marshy here and well worn, leading me first to the Geo of Sclaites, another deep chasm that dog-legs in from the coast and provides sanctuary to tens of thousands of seabirds, mainly guillemots and razorbills on the lower levels, with kittiwakes and herring gulls taking the top floors. They really were like floors here, for the rock had vertical striations that were spirit-level straight. The enclosed walls caused the birds' cries and calls to echo and ricochet, and the effect was not unlike a children's birthday party, albeit for children who would eat nothing but fish. The water was inky black but frequently spattered with white from the birds. Near where the geo met the coast, I saw a lone puffin parading back and forth on his own ledge, white chest proudly puffed out like a starched uniform shirt. There were fewer puffins here than at Dunnet Head, but then I was visiting in the middle of the day, so they may have been out at sea fishing. The best time to see puffins on land is just after dawn and just before dusk.

Great skuas patrolled the skies above the path, while my feet kicked through the remains of their latest victims, mainly dried-out rabbit, and kittiwake wings. Hardy

Scottish sheep munched loudly on the grass. A whimbrel scooted across the grass away from me. I crossed a couple of fields, using stiles, and tried to ignore the wind blowing in viciously from the sea. At least it wasn't raining, something always to be grateful for in Scotland.

After fifteen minutes or so, I reached the coast again. This time the view was dominated by the Stacks of Duncansby. These two formidable sea stacks rose out of the sea like sharpened pencils, as tall as the cliffs on which I was standing, about sixty metres up. The Grand Stack, which from a different view looked a tower of cards, appeared taller than even its neighbouring cliffs. Both were green with tufts of vegetation, around which lines of guillemots perched precariously. In some patches, there was enough ground for puffins to have made burrows. The tops of the stacks appeared block-like, as if made from LEGO. It's said that the stacks were earmarked for atomic tests back in the 1950s but the plans were scuppered by the Scottish weather, which scientists considered 'too wet'. Too dreich, as they would say around there.

As I walked further round, the rocky archway of Thirle Door came into view – a sea arch yet to become a full stack of its own, it was still tethered to the cliffs by its roof. We were truly into *Lord of the Rings* territory now. I was pleased the bombing had been thwarted: the stacks were truly stunning. The aquamarine sea around them was dotted with skerries, between which grey seals were wending their way to and fro. I could have stayed with this magnificent view for a good while longer. It reminded me of the Westfjords of Iceland – beautifully rugged and unkempt, and somewhere

Scotland should certainly be hugely proud of. I knew that the North Coast 500 had been trying to make the best of this under-visited corner, but this was something really special. As I got to the car park, a Dutch couple dismounted from a motorcycle and asked me to take a photo of them. I did so, of course, but then they just got back on the bike and left. No attempt to even get near the wonderful stacks, just the selfie shot and off again. It seemed such a shame. Their loss, I guess.

I sneaked back to Easter Head later on, at the end of a day when I'd been dogged by a low mood and a building sense of being under pressure. It was getting close to dusk, and the place was bathed in a blue light. Camper vans filled the car park, but their occupants seemed to have retired early. The RSPB shed had been secured to the ground using cables, a reminder that the weather there wasn't always so benign. The place was otherwise deserted, and I knew I'd have the clifftops to myself.

The lighthouse here was another of Robert Stevenson's – clearly a busy family, the Stevensons. The tall white tower stood proud of its low buildings; the glass dome ready to cast its light far out into the dark seas. It is now monitored from a remote base. I know that the life of lighthouse keepers was notoriously hard, but being monitored remotely takes some of the romance out of it. The lighthouse keepers here used to get occasional visits from the late Queen Mother, who once resided in the Castle of Mey, near the village of Dunnet. The Northern Lighthouse Board report that the keepers gave the Queen Mother tours of the lighthouse and a spot of afternoon tea, although the last time that happened was in 1979. I love the idea of lighthouse keepers sharing hot tea and cream scones with the Queen Mum while birds wheeled around the lighthouse and the sea pounded the base of the cliffs. I wonder what she made of the puffins, and what they made of her.

The lighthouse was surrounded by old military buildings that, somewhat eerily, had had their windows bricked up. The old farmhouse had been half renovated; exactly half was painted white and the rest was still crumbling away. Two families of lighthouse keepers used to live there. There was a patch of brand-new tarmac for no apparent reason. Three twite on a nearby fence watched my confusion, intently chattering away to themselves. I veered off the path to the north. The grass here was soft and springy like a wet carpet and I headed towards the clifftops once again. A blue boat was bobbing on the sea below and yellow-suited fishermen were pulling in blue lobster pots; the buoys on board looked like party balloons. Orkney felt so close, I could almost touch it, yet it was nearly seven miles away. I spotted a break in the rock where wildflowers had taken over; sea pinks and red campion were joined by sea mayweed. It was startlingly beautiful.

The cliffs there were high-rises for seabirds. Every available space was crammed with life. The acrid smell of guano and fish stuck in my nostrils, and the raucous chorus rang in my ears. This was truly the sound of summer to me now – the seabird orchestra in full swing. It was hard, and horrible, to imagine the cliffs silent and sad without their summer visitors. Impossible to contemplate.

Puffins ruled the top of the cliffs. I was pleased to see that they were there in great numbers. I spotted a couple billing, while another pulled at the grass. As the puffins passed by me in flight, I could hear their wings flapping. A puffin sat proudly on top of a tuft of sea pinks, striking the perfect pose for any tourist brochure. There was no shortage of birdlife here. I was stunned by how many puffins there were, but the general buzz of birdlife around the cliffs was equally staggering. On days like this, one could easily forget the plight of the puffin and its fellow seabirds.

I put Erland Cooper on, placed earphones into my ears, found a dry patch and lay back, feeling my stress levels start to drop. The puffins provided the perfect pick-me-up, while Erland's compositions soared and soothed me. There's a well-established link between nature and mental health, and I was only just beginning to feel the benefits. This book may have started with me trying to find the puffins before it's too late, but it was becoming clear that they were helping me too. I could feel the stresses and strains of life starting to dissolve. The puffin pulled at another blade of grass, twisting his head sideways to consider me fully.

11

Puffins Rule

June

Sumburgh Head, Shetland, Scotland

It had not been a pleasant day. The series of buses, trains
and planes required to get me to Shetland had taken its toll,
propelling me through several hundred miles of a damp
blanket of grey cloud that was smothering the whole of
the United Kingdom, even up to this northern outpost. At
least the Loganair flight from Aberdeen had been full of
Scottish optimism, with the pilot calling some quite severe
turbulence 'a wee bit of bounce', and those Tunnock's
caramel bars you only find in Scotland.

I stepped off the plane at Sumburgh, at the southern
end of Shetland's Mainland Island, into persistent drizzle and
low cloud. This wasn't ideal, as the ground crew had decided
to deposit our cases – somewhat unusually, I felt – directly

at the bottom of the aircraft steps, on the wet tarmac, in the rain. My mood was dampened further. I disentangled my bag from those of the other passengers and trudged into the airport building. This was clearly under renovation, encased in wooden boards and scaffolding. There were more fluorescent-clad workmen than passengers in the terminal. The place was cold, and smelt of sawn wood and aircraft fuel. I decided not to linger, so passed through the annoyingly slow revolving doors in a shuffle, to a waiting bus.

'Are you going to Lerwick, mate?' I enquired of the driver through the open doors.

'Yeah, once I've had my break,' he said, promptly closing the doors in my face.

I stepped backwards to save getting my nose caught, and into a large puddle. I felt cold Scottish water leaking into my warm trainers. I won't repeat here exactly what I said, in case there are children reading. I cursed the weather, the UK's inefficient transport systems, rude bus drivers who favoured ham and piccalilli sandwiches over customer service, and the daft idea of trying to visit puffin colonies in the far north of Scotland.

It was not my first trip to Shetland. I'd been before, although I was about ten at the time and was with my local Scout pack. I don't remember much about that trip, except staying in Lerwick, the complete lack of trees anywhere on the island, team-building sessions involving homemade rafts, and an abject sense of homesickness that peculiarly focused on the boiled egg and toast soldiers of home. I've had a soft spot for boiled eggs ever since, but Shetland failed to make an impression on me.

I got a more hospitable response when I phoned the Sumburgh Hotel, which kindly sent out a driver to collect me from the airport. It was a rather old-fashioned hotel, fashioned out of austere local stone and perched at the end of the airport runway but within sight of Sumburgh Head, which rose out of the sea in the distance, its impressive lighthouse taking centre stage. Outside, guests were greeted by two genuine and extremely cute Shetland ponies, both of which looked like they had borrowed their hair from a 1980s pop star. The hotel itself was all striped wallpaper, red carpets and ringing bells and had an elaborate system for booking evening meals. The staff were extremely friendly and spoke in a rapid Shetland accent; I particularly enjoyed the greeting of 'aye, aye' in the mornings rather than a boring 'hello'. I considered adopting that myself.

I asked the friendly girl on reception how long it would take to walk to the lighthouse. She gave a reply that I didn't quite catch, but I was too shy to ask her to repeat herself. It ended in 'something like thirty minutes' anyway. I decided to give it a shot, despite the receptionist looking at me like I had lost my mind with the rain lashing hard at the window. 'Welcome to Shetland,' she said with a half-smile as I stepped towards the door.

The wind was constant. It didn't blow in gusts, like normal wind, but in a continual stream that occasionally strengthened to the point that I worried my six-foot frame would lose contact with the ground. The rain was squally, and had started to ease off, although the wind ensured any remaining drips were thrown into my face with force. My eyes were streaming as I meandered through a farmyard of the most depressing farm ever, with its greening asbestos roof, brown square walls, tiny slit-like windows and peeling paintwork. I picked my way over abandoned fence posts and a path almost entirely covered in sheep excrement. Once through a series of gates and stiles, I gratefully made my way onto a newish single-track lane of clean black tarmac that weaved its way through tight grass fields and neat stone walls to the lighthouse in the distance. The stone walls were topped with tufts of green lichen, a testament to the damp environment and the fresh, clean air. I passed a defunct quarry, its scars showing to the world and now home to several pairs of kittiwakes, which were hunkered down deep into its crevices against the wind and rain. Nettles grew at the base of the cliff, amongst long-forgotten tyres and other farming equipment left to rust. I continued on.

I approached the lower lighthouse at the bottom of the road. I was frequently overtaken by motorcyclists covered in leather and spray from the road, and, oddly, a coach trip of elderly folk all the way from Bath. This smaller lighthouse had been moved there from Muckle Roe in the west of Shetland in 2001. It was clearly well looked after, in bright, resplendent white, one of thirty-nine lighthouses on Shetland. That's a lot of white paint. The guidebooks told me that if I looked over the car park wall, near the Slithers, I would see puffins. They are the most accessible puffins in the UK, the guidebooks said. It turned out that they were entirely wrong. Not a single puffin could be seen on the steep slopes leading into the gloomy seascape. I hoped this wouldn't be the case for my whole trip.

I read the plethora of information boards that grew like plants there, but the drizzle was coming again and I could feel it dripping from my nose, so I pushed on. It was a steep climb from there on, a real head-down job. I passed a chunk of whale vertebrae that had now got its own information board, and, surreally, a full-sized plastic sculpture of an orca. It felt quite out of place in that rugged, natural landscape and should obviously have been in a theme park somewhere. Very odd. Near the cattle grid, about halfway up the road, was a small statue of a trow, a kind of local troll.

Turning left, a rocky outcrop led to a sharp gully. This was an amazing spot. At the bottom, the sea was in turmoil, churning over and over in whites and blues. The grey rock was exposed in patches but, on both sides, the gully was covered in pink thrift, or sea pinks, as they're known there. It was as if the whole place had been attacked by a young child with a pot of pink paint. Between the patches of sea pinks I could see hundreds of burrow entrances. This was prime puffin territory. There were a few puffins out, but not many, and I couldn't blame them in that weather, when the warmth of the burrow or the necessity to find fish out at sea must have been a priority. At the bottom, a colony of guillemots noisily protested against the waves and wind, scolding the elements with their constant chatter. I made a mental note to return to this spot. Rather wonderfully, there was a sign that read 'For the Safety of Researchers, Please Do Not Disturb Them'. I hoped it was for the puffins to read, and I amused myself with this thought.

A few more twists of the road led to the summit and a large collection of low-level buildings towered over by the main lighthouse. Next to it stood a building carrying the most wonderful foghorn, painted in a splendid pillar-box red. Beneath it were two equally red tanks, from which the air was pumped. On special occasions, the foghorn is sounded, presumably to the great surprise of the puffins over the wall.

Three flags were flying bravely in the wind – the Union flag, the Scottish flag, and the white Nordic cross on blue for Shetland. All three were in danger of being torn apart in the wind. Near the flagpoles was another chunk of whale, this time a skull from a minke whale. I went and took a look, only to surprise a rabbit which was taking shelter in one of the eye sockets.

The lighthouse was impressive – whitewashed and with a cylindrical, crisscrossed glass top disguising the powerful light within. A couple of the buildings had been used for radar during the Second World War, and there were a few exhibits that cost a rather pricey £6.50 to view. The RSPB office appeared to be unstaffed, and the tiny shop had an unsmiling girl behind the till who insisted on chasing anyone who'd wandered inadvertently out of the shop area and into one of the exhibits for a ticket.

There was a 'pop-up' café a bit further down. Why is everything 'pop-up' these days? The only thing that used to be 'pop-up' in my day were books. It was in a wonderful building though, right on the west side of Sumburgh Head, perched on a clifftop and with a crescent of large windows that allowed 180-degree views across the southern half of Shetland's Mainland. I could see, through the murk, the airport with its orange lights and several red RAF helicopters waiting to lift off and, further away, the rising hulk of Scatness with Horse Island beneath it.

I had wanted a greasy bacon sandwich and a cup of builder's tea as ballast against the wind for the walk back, but it wasn't that sort of café. I settled for a gluten-free chocolate brownie that was surprisingly good, and a dainty cup of English Breakfast tea, which felt entirely wrong, especially as I was sharing the café with a group of sodden, leather-clad, neckerchief-wearing bikers. No one cared though; they were all transfixed by the view and the birds soaring around, despite the wind. A gasp went up from the patrons of the café, and there was a flocking to the windows. To everyone's astonishment, a huge great black-backed gull flew directly past the glass, with a brown chick in its beak. Lunch was served. 'In-flight meals' took on another meaning.

Rather than following the road back to the hotel, I took the West Voe path, which twisted its way along the coast and caused me to cross some high and precarious stiles. It was worth it though. I spotted a few puffins from a distance, but I had other company. An oystercatcher peeped at me from a fencepost, and a wheatear followed me for a good distance. She fluttered past and then sat whistling and rattling to me. As she flew, she showed flashes of white; she may well have been trying to distract me from a nearby clutch of eggs.

As I got closer to the hotel, the stone walls were splotched with the egg-yolk yellow of another kind of lichen, and the unshorn sheep were taking shelter behind them. In the distance, I had my first sighting of a great skua, or bonxie. It was chasing an oystercatcher around, but the oystercatcher was more than capable of giving him the slip. The fight slipped over a wall and out of my view. Back at the hotel, I spotted a tiny Shetland wren flitting from stone to stone in the wall; his loud, operatic singing was giving him away, and his tail twitched back and forth. Outside, the Shetland ponies were talking to each other about the wind, and hairstyles.

Later that week, I was searching for another bird for a change – the elusive storm petrel. At Cunningsburgh, I boarded the boat to the island of Mousa in the dusky half-light and with some trepidation. The sky was pink with the dipping sun and it felt strange to be taking a boat so late in the day. The boat was half covered with a tarpaulin roof, but the back was open to the elements and I wasn't accustomed to sailing at night. The sea was calm though, and I tried to relax as I handed over a crisp Scottish tenner to the boatmen who had promised to take us to find storm petrels. That wasn't the only cause of my unease though; some sailors consider storm petrels to be cursed, the bringers of bad weather, or, worse, the spirits of sailors who have drowned. I tried to push such thoughts out of my head, instead busying myself with making puns out of 'petrel' and 'petrol' stations. I nearly made myself smile.

Storm petrels are no bigger than most garden birds, which makes them relatively tiny for a seabird and quite amazing considering the rough seas and atrocious weather they must contend with on their annual migration to South Africa. When searching for food, storm petrels dangle their strange long legs above the sea with their feet almost dancing across the surface, all the while holding their wings upwards as if worried about getting them wet. They are a special little bird.

It's their odd-shaped legs that prevent them from walking properly on land too. They have an excellent sense of smell and, unusually, have their own strong scent, which helps them find their nests in the dark. Storm petrels do quite a lot in the dark, mainly to avoid predation, and could easily be mistaken for bats in flight.

'Mousa' comes from the word 'mossy' and is pronounced 'Moosa'. Mousa is best known for its Iron Age broch, and that's where we headed as soon as we docked. It was a steepish climb and the grass was littered with the remains of sea urchins and crab claws from the great black-backed gulls that roam the island in the daytime. As we walked along a stone wall towards the broch, I heard my first storm petrel. It was the strangest noise. One of the boat crew laughingly described it as the sound of a fairy throwing up, and that's a pretty good description. In the dark, it was almost eerie; no wonder storm petrels used to be seen as harbingers of darker worlds. The call was like a high-pitched purring, with a short gap, a click, and repeated over and over. I couldn't quite make out where the petrel was, but quite clearly it found the wall a suitable place to nest.

The broch was just over thirteen metres tall and made entirely from stone. The best way I can describe it is windmill shaped, but without the sails, obviously. Some two thousand years old, it's thought to have been built for defence purposes. I was keen to explore it. It had a tiny door at the bottom, facing the sea, through which I could just about fit, mainly by uncoupling several of the vertebrae in my lower back. Inside, there was a stone floor and fireplace and then another small doorway that led to an extremely narrow staircase, which wound between the thick walls to the top. It was a one-in, one-out type deal and I waited my turn.

The storm petrels had yet to show in numbers, but someone had produced a stuffed one, and it was being passed around. Almost entirely dressed in black, they really are quite small and the model sat easily in the palm of my hand. These delicate little birds only return to land to breed and lay one tiny egg per year. They visit their offspring only at night, regurgitating fishy food to sustain them.

By the time I returned outside, the broch had swirls of petrels flying around it in the half-light, each letting out an otherworldly call to its mate. More and more birds joined the spiral, until the light gave out almost completely. All I could do was stand and stare at this phenomenon. I'd never seen or heard anything quite like it.

The wind was still raging when I returned to Sumburgh Head a few nights later, and I wondered if there was ever a time on Shetland when it wasn't. The complete lack of trees or cover didn't help and, when I reached the clifftop, the wind was beating my ears like a bass drum and making the straining wires sing. If only the foghorn had sounded, we would have had a complete band.

It was late evening and I guessed there wouldn't be many puffins about. I feared that the strong wind would have caused them to dive for cover and, in any case, chicks were hatching underground and needed feeding. Pufflings need constant parental attention for the seven or so weeks it takes them to fledge. They'll stay in the burrow for most of this time, getting closer to the entrance to meet parents with food as they get older and braver. Once ready to leave the burrow, they do so alone, heading out to sea without parents, already pre-programmed with what to do and where to go.

The first sign that I was wrong were the photographers who were out in force. There were several hanging over the eastern wall, while the slopes below the observatory were also busy. The café in the observatory had closed for the day – what's the opposite of pop-up? Fold-down? – and, instead, a lone figure stood at an easel, the concentration etched on his face as he tried to render the busy landscape onto his canvas.

I walked down the wall on the eastern side, its whitewash starting to crack, revealing its sombre grey interior. Rabbits scuttled from under my feet. I looked over the wall at the steep cliff face, bisected by a brief gully. The sea was heaving and shoving at the rocks beneath me. I could taste salt on the air. The gully was covered in sea pinks, but well-worn paths gave away the locations of the burrows. I stopped and watched for a while. Just down from the wall, a sight stopped me in my tracks. A dead puffin. Face down in the grass, with wings tucked under the body, it looked like a murder victim in a B-movie. The usually pristine black back feathers were bedraggled and dishevelled, and I could see pale grey skin beneath. I couldn't see any feet. This puffin had clearly been predated; if it was flipped over, I'd no doubt see that the thick chest meat had been removed, probably by a great black-backed gull.

Back below the observatory, yet more puffins had appeared. They were sprinkled liberally across the slope. There was more grass here, but they seemed unconcerned. I found myself a nice little place next to a metal gate that was bolted and welded shut and sat and watched for a while. The sun was trying to break through the clouds, but the grass was still wet from earlier showers. The photographers began to thin out, and I saw the cars leaving the car park one by one. I couldn't wait until they had all gone, until it was just me, the howling wind and several hundred puffins.

I spotted a puffin sitting near the wall, and moved across to see it more clearly. It wasn't bothered by my presence, but just sat carefully cleaning itself, using its beak to check and inspect each feather in turn. I manoeuvred a little closer, in the way that only a man over six feet tall and the wrong side of thirty-five can do – inelegantly, and with lots of huffing and puffing. I leant over the wall while trying to keep a low profile against the sky. The wall was rough and it grated against my soft, slightly flabby skin. I may have sworn. I was probably some fifteen metres away, and the puffin was still entirely unperturbed. It was at this point that I noticed the camera next to me. Rising from the ground was a stick, and on top of that sat a rotating globe containing a camera. My attempts at getting a better view of this puffin had been beamed live across the internet. To anyone out there who was watching, and had their evening spoilt by an Englishman cursing to himself whilst leaning over a wall in Shetland, I apologise. I seek your forgiveness. I looked again at the puffin. I could have sworn he was smiling gently to himself.

Red-faced, I walked back up towards the main lighthouse, passing a couple of tumbledown buildings that had long since lost their roofs to the wind and rain. As I eased myself through the five-bar gate, I saw some graffiti on one of the doors. In thick black pen, someone had drawn a rudimentary puffin – it looked a lot like a pheasant – next to the legend 'Puffins Rule'. I'm not a huge fan of graffiti, but this made me smile. Puffins do indeed rule.

Further down, I leant over the wall and watched the puffins going about their evening's business. More and more had appeared, either from out at sea or from burrows. The slopes were now dotted with hundreds and hundreds of the little humbugs, their sugar-white chests standing out against the grass, each one facing out to sea. I was particularly enamoured of a pair nearby, who were acting like, well, lovebirds. They had a burrow perched neatly between two perfectly manicured domes of thrift, from which they kept emerging, rubbing bills and snuggling up to each other. It was as if they were kissing and cuddling. After reaching maturity at around three or four years old, puffins return to the same mate year after year, but I guess there is always a first year. This certainly appeared to be new love, a honeymoon period. They sat side by side, the wind ruffling their feathers. Occasionally, one would return to the burrow as if to check on something, before coming back out onto their well-tended patio.

Back below the foghorn, I discovered that the photographers had packed up and cleared off. I had the place to myself. I walked up the slatted wooden walkway to peer over the wall and immediately saw what had got the photographers frothing at the gills with excitement. The wind here was so powerful that the puffins on the very top of the cliff were being forced to carry out difficult aerial manoeuvres just to land. They could take off just fine, but landing demanded skill and effort, often involving many minutes of being stuck just a couple of feet up in the air, with a worried look on their faces. Wings were furiously flapped and orange feet outstretched as a steering mechanism. A few were using the wind to land backwards, tails first, but this required peering around their own bodies to get a view of the landing patch. A set of mirrors was evidently required. Puffins failing to master this technique were thrown to the ground with a painful-sounding thump, resulting in a confused, unusually ruffled bird having to take a few moments to get its bearings – unless it landed on another puffin, in which case all hell broke loose. I stayed to watch this spectacle for a while. It reminded me of the Red Bull Air Race, where pilots pull stunts in impossibly tiny planes over the River Avon in Bristol. They have worried looks on their faces too, and I'm sure there have been moments of feet flapping.

Down at the gully next to the orca – Orca Gully? – the wind wasn't quite so strong. I sat on the grass and just watched the puffins come and go. My right foot was next to the low wire fence designed to keep puffins away from researchers. Through my foot I felt an odd vibration coming from the ground. About thirty centimetres away, near the fencepost, was the epicentre of this sensation. It was a puffin. It popped up, spread its wings and looked directly at me, not giving a damn. I hadn't been this close to a puffin since

Staffa, and those puffins had been busy, hurried. This chap was content, relaxed. His beak was beautifully coloured – a sign of a healthy bird – and I was so close, I could even see his gorgeous orange eye make-up too. The light was going now, and it was close to 11pm. I reached into my bag and pulled out a bottle of local beer. Simmer Dim from Lerwick Brewery. The puffin and I gazed out to sea. Puffins don't drink beer, but I raised a toast to him and his Shetlandic puffin friends.

12

The Final Checkout

July

Unst, Shetland, Scotland

There's something strange about Unst, the furthest north of the Shetland Islands. It is as far north as you can go in the UK, on the same latitude as southern Greenland and seventy-five miles away from Sumburgh. I felt it as soon as the ferry docked at Belmont on Unst's southern tip – this was no normal place. It was the opposite of its southerly neighbour, Yell, which had seemed dank and dour. Yell is flat, and its only claim to fame seems to be that it has the highest concentration of otters in Shetland. Luckily, I had spent no more than twenty-five minutes there, which was the precise amount of time it took to whizz from the very southern end of Yell – the port at Ulsta, where ferries from Shetland's Mainland dock – to the port at the northern end, for connections across the Bluemull Sound to Unst. As soon as the ferry pulled in at Ulsta, engines were revved and we were off, tearing across Yell in a mad Wacky Races, through the fog to Gutcher, eighteen miles away. Yell's sparsely populated interior was dotted with low, grey farmhouses that looked like something out of a horror story. You know, the ones where they drag you in and do unspeakable things to you with surgical implements.

Unst had a much happier feeling about it from the start. On arrival, there was no rush – there was nowhere else to get to, that was for sure – and it was all a lot more welcoming. The first thing I spotted was a three-foot-tall puffin statue celebrating the fiftieth anniversary of the moon landings, one of several across the island, each dedicated to a different astronaut. Next up, there was a sign for a Chinese takeaway in Uyeasound in the south of the island. Nothing odd about that, right? It was for Sunday evenings only, 4pm to 7pm. It gave me a hankering for Singapore noodles, but I'd just have to wait until Sunday.

From Uyeasound, I made my way north to Baltasound, passing Loch of Watlee, where a curlew was making an ungainly landing and a red-throated diver glided effortlessly through the still waters. Nearly a third of the world's population of red-throated divers breed in Shetland, where they are known as rain gös – rain geese. It's said that they can help you predict when it's going to rain, although in Shetland that seems to be much of the time.

The madness continued, first, in the shape of a red double-decker bus parked up on the side of the road; I had no idea how it could have got there as it certainly wouldn't have fit on the ferry. Then came a six-foot-tall figure made out of two rocks placed on top of one another, wearing a pair of swimming goggles. The radio was full of local news – a Land Rover had been found with a damaged wing mirror, and there'd been a fire, but the Fire and Rescue Service had dealt with it by 'inspection only'. I liked this place already, and I felt the ghosts of Yell falling away.

Formerly a busy herring fishing port with a population of almost ten thousand, Baltasound was now just a few houses alongside the road, and a couple of shops. Everything here was the 'most northerly' – post office, lottery, church. My favourite was the Final Checkout, a petrol station and café that sold everything you could ever need and was clearly the centre of the community. There was internet access, a communal guitar, and local artwork for sale, and you had to remember the price of the petrol for the girl behind the till. It was that sort of place. Locals with hangovers (Unst has its own distillery and brewery at Saxa Vord, a depressing-looking former RAF base that's been trying to regenerate itself since the base closed in 2006) discussed the events of the previous night, while tourists asked for everything from lighter fuel to nail clippers. They had both. I plumped for a lava-hot cheese and onion toastie and a bucket of scalding coffee while I absorbed everything that was going on around me. It appeared that more gossip than money was being exchanged, but no one seemed to mind.

At the junction where the road forked to the Keen of Hamar Nature Reserve, I spotted another little Unst anomaly. Bobby's Bus Shelter. The story goes that a seven-year-old Bobby Macaulay used to use the bus shelter to wait for the bus to school each morning. Given how variable Unst weather can be – I'd already had fog, sun and rain that day – I could see how essential this would be. So, when Bobby found that Shetland Council had removed the shelter without warning, he wrote a letter letting them know of his disappointment. It worked, and not only did the council replace the bus stop, they furnished it for Bobby too. That was some years ago, but the tradition has been kept up – I found the little red shelter had been carpeted and contained a microwave, phone, chair and table, a Sully from *Monsters Inc.* cuddly toy, curtains and a library of books. Outside, another stone puffin stood guard, this one dedicated to Galileo Galilei. Both the puffin and the shelter made me smile.

The Keen of Hamar, by the way, is the only place in the world where Edmondston's chickweed grows – a small plant with oversized delicate white-and-purple flowers that has adapted to growing just in this one place. The chickweed was found by Thomas Edmondston, a local botanist, at just twelve years of age, hence its name, although I think I prefer its alternative name of Shetland mouse-ear. Edmondston unfortunately met his end at an early age, not on the Keen of Hamar searching for plants with mouse ears, but accidentally being shot in the head in Ecuador.

The main road that bisects the island took me to Haroldswick, which had a replica Viking longhouse and ship by the roadside. I explored both the ship and the house, watched with bemusement by a couple of Shetland ponies over the fence, shaggy forelocks over curious eyes.

There were further boats on display at the nearby Unst Boat Haven, which was a handy place to hide from the rain while learning about Unst's maritime past. Most of the twenty or so salvaged wooden boats dated back to the herring boom at the tail end of the eighteenth century. My favourites were the deliciously named *Laughing Water*, and a bizarre prototype of a folding boat. There was also the inescapable feeling of being in your grandad's shed, among lots of fascinating bits and pieces kept 'just in case' and with nowhere else to go.

I headed further north towards Hermaness Nature Reserve, along single-track lanes with splashes of yellow, pink and white flowers. Summer had arrived and its graffiti was all over the verges. Five great skuas or bonxies were lined up on the shore of Loch of Cliff. Unst has a substantial population of great skuas and it seemed they favoured this half of the island. I silently thanked Dr Laurence Edmondston, father of poor Thomas, for having set up a watch scheme in the 1800s to protect great skua eggs from the then popular hobby of egg collecting.

I passed the stunning little beach at Burrafirth, a swathe of brilliant yellow sand set against the blue sky in what could easily have been a corner of Scandinavia and, beyond that, several sea stacks with awesome names like Boo Stacks and Burgar Stacks, before eventually coming to the clutch of low white buildings comprising Muckle Flugga Shore Station. This was once used by keepers from the lighthouse precariously perched on Muckle Flugga, who took it in turns to stay at the shore station or at Muckle Flugga. It was now meant to serve as holiday accommodation and as the visitor centre for Hermaness, although the latter was closed for the year, no reason given. There was a sundial with a Shetland pony ornament atop, a helipad amongst lush, green grass and a fantastic view out to Burrafirth. Tempted as I was to find a spot and eat the sandwiches I'd squirrelled away from the Final Checkout, I needed to get up to Hermaness itself.

The car park at Hermaness had warning signs about fog, strong winds, wearing waterproof trousers (you were more likely to slide off a cliff), great skuas, red-throated divers (who did not attack but were protected – even from photography in the breeding season), dogs, plants, and litter. Undeterred, I pushed on up the steep path to Mouslee Hill, which was lined with gentle purple flowers known as lady's smock. The path was well defined and soon became a boardwalk, presumably to protect the sphagnum moss and peat beneath. I was grateful, as the ground was evidently sopping wet; perhaps not surprising given that the path followed the burn of Winnaswarta Dale. I could hear water burbling beneath me and, occasionally, squelches under the boardwalk, reminding me of the fairground water flumes of my childhood. A common frog surprised me as he leapt across the mounds of moss; I don't know why, but I hadn't expected to see one this far north. He grinned at me with a wide mouth and shiny eyes, knowing that he'd been seen. I left him alone and moved on.

I suspected that most visitors dashed across the boardwalk to get to the cliffs and the bucket-list tick of the other side. They'd be missing out though. Great skuas nested amongst soft moss, and white bog cotton and yellow tormentil punctuated the many shades of green. Even with a cold fog, or haar, as Shetlanders call it, starting to come in, it was surprisingly beautiful. The sea fog lent an additional calmness, and even the great skuas didn't fuss as I passed. One was gently bathing, showing off its strong body with white flashes on outstretched wings, reminding me of police sergeants' epaulettes.

The haar got thicker and thicker as I approached the sea cliffs, to the point where I couldn't even see a few metres in front of me. Just when I needed it most, the boardwalk stopped and I was left following a series of wooden posts. I was glad when I reached the cliffs; from here, at least, it was simply a matter of walking north, providing that I didn't step off the cliffs in the thickest fog I'd ever (not) seen, and that I could find the boardwalk again on my return. I considered leaving a trail of breadcrumbs.

As I walked along the clifftop, lamenting having come this far only to see nothing but my feet, I could feel the haar soaking my skin and clothes. It was cold and it would have been easy to turn around and head back to the enticements of the car heater and my butties. It was then that I realised that I was right on the edge of the cliff and in danger of kicking puffins. You read that correctly. They were so close, and I couldn't see them until I was almost treading on them. I made out a vague silhouette in the peasouper and then I was inches away from a Hermaness puffin. Despite the cold, I was smiling like it was Christmas morning; each puffin was a new gift under the tree. They were unconcerned by my presence, but I was keen not to disturb them and took a few steps away. I was afraid of treading on burrows. I also had little faith that they'd call the emergency services if

I did take a tumble over the side. I was thrilled to see them though. I even got to see a puffling – freshly emerged from the burrow, with some fluffiness left, dark cheeks, and a black, pointed beak. It was July, so they'd be off out to sea soon.

With the haar, I guessed that the bonxies and gulls couldn't see anything much either, so the puffins would have felt a little safer. The only sounds I could hear were the low growls and chuckles of the puffins as they talked to each other. I'm always amazed that this sound comes from puffins. It doesn't seem to fit at all – a cartoon with the wrong soundtrack.

I continued on the path, leaping over West Sothers Dale and then East Sothers Dale while trying to keep my feet dry. The haar started to clear and I could see that I was actually surrounded by puffins; they were everywhere. On a single boulder, I counted a dozen puffins seemingly having a board meeting. It struck me that there may have been more on that boulder than on the entire coast of mainland Devon and Cornwall. On a recent trip to Cornwall, I had found only a single puffin off the Mouls in Padstow. A sad state of affairs.

The curtain slowly lifted to reveal a tremendous vista – the sea foamed white at the base of the cliffs beneath me, and the whole was framed in that delicate blue light you get only in the far north. Sea stacks dotted the coast and, to the south, I could see a beautiful cove sprinkled with boulders beneath tall cliffs. Puffins flew back and forth – a skyway full of birds. The sun began to shine, burning away the remainder of the fog and unveiling to the north a series of impressive stacks – Flodda, Urda, Clingra, Humla, the Greing, and a resplendent sea arch. And then, just peeping over Hermaness, I spotted the top of Muckle Flugga lighthouse.

As I got closer, I could see the gannet colony. It had stained the clifftops and stacks bright white, just the same as the lighthouse paint. Thousands of huge solan geese, as gannets are known on Unst, circled in the sky above and beneath me, before plummeting into the sea, black-tipped wings folded back behind them and specially strengthened creamy yellow necks (think Cornish clotted cream) outstretched. They have built-in air pockets to cushion against the impact of hitting the sea repeatedly, although they rarely live past the age of six – their diving habit frequently causes blindness or a broken neck. Others sat facing the cliff face on nests made from seaweed and glued together with their own guano, each equally spaced from the next. The ultimate in social distancing.

While I sat and observed this raucous seabird festival, I was reminded of the story of Albert. At the Saito gannetry, about a mile further south, there was a celebrity of a bird called Albert, a desperately lost black-browed albatross. These odd-looking birds, named for their mascara-like facial markings, usually reside in the southern hemisphere, some way away from Shetland. In the 1960s, however, Albert was spotted at the gannet stronghold at Bass Rock off the Isle of May. After this, he visited Hermaness nearly every summer for over twenty-five years, much to the amusement of visitors and birdwatchers alike. It's thought that he was looking for a mate, for he built a nest each season, although sadly he was destined to remain single. Albert disappeared from Hermaness just as quickly as he appeared, though there were possible sightings at Sula Sgeir, forty miles north of the Isle of Lewis (another favourite of gannets) in the late 2000s. It's possible this was Albert, as albatrosses have life expectancies of up to a staggering seventy years. In 2020, a black-browed albatross was seen at Bempton in Yorkshire, although this was thought not to have been Albert. Let's hope Albert found his way home, even if only spiritually.

Twenty-six thousand pairs of gannets make a substantial racket with all their gurgling and gargling, and with the smell too – it was quite a hit for the senses. A hundred years ago, not even a handful of gannets were counted on these cliffs, but now they were thriving, their numbers increasing at around four per cent each year according to Scottish Natural Heritage, which is a phenomenal rate. It must be really tricky to count the noisy, quarrelsome birds these days. As I looked out to Muckle Flugga, about a mile and a half from the edge of Unst, I noticed that the gannets had painted the Rumblings stack white with guano too. That's nearly as far as you can go in the UK and, brilliantly, it is owned and dominated by seabird colonies, that were quite literally making their mark.

Figures from Scottish Natural Heritage also show that there are somewhere between twenty thousand and thirty thousand pairs of puffins on Hermaness, a sizeable colony. They too were spread around the clifftops, despite the fearsome descent to the sea beneath. The general consensus is that they are in decline here, although on days like today that could be difficult to believe. Right now, the puffins were busy fetching fish from the sea, returning with mouthfuls of shimmering, silvery sand eels. It would have made a classic puffin photo but was tricky to catch. After landing inelegantly with wings and orange feet outstretched, the puffins didn't linger for a photo op but, instead, dashed down into their burrows to feed their chicks beneath. I also noticed that while I was trying to catch the perfect photo – and please bear in mind that I am by no means a photographer, I merely play at it – another puffin had wandered up to me as I sat on the cold grass and was less than a metre away. Not wanting to disturb the bird, I remained perfectly still while we checked each other out. He with careful twitches of the head and I with the widest of eyes. I could clearly discern the yellow rosettes on either side of his beak, and the 'stitches', where it looked uncannily like his beak had been sewn on.

Through the plunging gannets and the last wisps of the haar, I could see Muckle Flugga and its lighthouse and, beyond that, Out Stack. Muckle Flugga lighthouse was previously known as North Unst lighthouse and it sits precariously yet bravely on top of its outcrop of rock. No one thought that it would be possible to build a lighthouse out there – hence the nickname 'the impossible lighthouse'.

Lighthouse keepers were stationed – in teams of three – on Muckle Flugga until 1995, when it became automated. It must have been a difficult and lonely posting. From written accounts at the Unst Heritage Centre, I learnt that the keepers did what they could to survive physically and mentally. Library books were taken out via the boats from the shore station at Burrafirth, one keeper could play the fiddle (although I assume this would have grated on colleagues after a while), while another began to sew clothes for his wife. Writing, painting and woodcarving seemed popular, and communication with Burrafirth was via a series of coloured flags. Hens were taken to provide eggs, in addition to those of seabirds, and there was a photograph of a keeper hugging a seal that made me chuckle. Neither keeper nor seal looked like they wanted to be there, on a bitterly cold and dangerous rock in fearsome seas at the end of the world.

At the end of my stay on the island, as I waited for the ferry back to Yell from Belmont, I went for a walk, hoping to find an otter. Unst hadn't finished with me yet.

Otters have always featured in my life, growing up as I did near a river in rural Shropshire. Sightings as a child were rare, but I'm glad that my own children take great delight in heading down to the river on hot summer evenings to try and glimpse an otter playing. Sightings are still infrequent but are met with enough awe to cause momentary silence except for the sound of ice cream melting from cones.

Otters have done exceedingly well in Shetland. They are not, as some people will tell you, sea otters but otters that are normally found in freshwater elsewhere, now successfully adapted to marine life in Shetland. It's not uncommon to find them chewing on the type of crustaceans and shellfish usually encountered only in high-end restaurants. They have to bathe in freshwater at least once a day to wash out some of the salt and ensure waterproofing and warmth, and a local had told me that otter bath time on Unst was usually between 4pm and 6pm. I checked my watch – I was in luck.

Otters are called dratsies in the local dialect. Spotting them is far from easy until you've got your eye in. The same local told me that otters often walk through Lerwick harbour undetected; they are that sneaky. One had been known to enter the fish market and steal expensive cod and halibut from beneath the fishermen's noses. A Shetlander I met on the beautiful golden sands of Northwick beach in the north of Unst told me that she found footprints of otters on the beach every day but never actually set eyes on them.

Despite their ability to disappear, otters are thriving in Shetland and the population has actually increased to around one thousand. They used to be hunted here – often using stone traps into which they were enticed – but no more. Much more recently, otters also suffered through the use of now banned pesticides and the pollution of watercourses. These days, otters are much more likely to be sought out by tourists and photographers alike, and I can't blame them. Otters are the new celebrities of Shetland. Call it the *Springwatch* effect.

I kept my eyes peeled as I searched for otters on the shoreline. Everything seemed to turn into an otter – rocks, crows, folds of kelp. The sea was calm, gently lapping at the shore. I found a dilapidated pier and decided it looked decidedly otterish, so I settled down for the otters to come by. They didn't, of course, and after an hour or so I was getting uncomfortable. I thought about heading back and swung my rucksack over my shoulder. Feeling slightly melancholic

– my trip to Shetland was coming to an end – I started my amble back to the rental car. As I did, I saw some something coming towards me. It couldn't be, could it?

It was. It was a dog otter, and a big one at that. He was heading straight for me, coffee-coloured fur slicked and spiked, with creamier fur on his belly and cheeks. I could see his whiskers clearly, and his bright, glistening eyes. I stood stock-still as he lolloped towards me. Even on land, an otter's movement is fluid, rolling like the water in which they spend so much time. I didn't have time to reach for the camera on my back and, to be honest, I didn't want to. I tried to imprint the moment on my mind instead, while a wide smile spread across my face. The otter didn't even look at me, but headed into some long, wet grass, between yellow marsh marigolds and on down to the shore, where he entered the water without a sound, his long, smooth tail the last thing to slide beneath the surface.

I resumed my seat, to see if he would remerge. Otters can only spend a couple of minutes underwater before having to come up for air. It was just a question of where. I tracked the stream of bubbles until they disappeared from view. I'd be missing that ferry. There would be others.

13

Unexpectedly Handa

July

Handa, Sutherland, Scotland

I wound my way up the stunning west coast of Scotland, through deserted Sutherland from Ullapool – I'd always wondered where Ullapool was – to Tarbet. I stopped briefly at the Spar shop in Scourie. It felt like the last shop on Earth, and I bought an extra bottle of Lucozade, just in case it was.

Tarbet is approached by a series of red-sanded, single-track lanes that meander through glens and handsome lochs and eventually past a beautiful freshwater pond, complete with white lily flowers and pads. Tarbet itself was just a few buildings but included public toilets, a restaurant/café and a couple of houses set around a really pretty inlet. On the shore was a shed, a proper 'man shed' – it had its own kettle – which served as the ticket office for the Handa ferry. A colourful array of ropes, buoys and lobster pots lay on the floor, and a life ring was tethered to the front wall. I bought a ticket from one of the two men enjoying the shed's facilities and was told to wait a few minutes until more people arrived. This seemed extraordinarily optimistic as I was the only living soul other than Steptoe and son in the whole of Tarbet. Handa hadn't been on my original list, but a cancelled trip out to St Kilda meant a hurried change of plan. I knew that Handa had a few puffins but had seen a serious decline in seabird numbers in the past decade. I hoped there would still be some left for me to admire.

I took a seat on a mound of grass and reminded myself that there was often a different, slower pace in such places. Out to sea, six or seven small skerries hid Handa from view. Small white and yellow boats bobbed gently on the waves, and purple heather clung to the harbourside cliff face to my left. A shallow barge had been pressed into service as a float for lobster pots, which the swell was constantly threatening to topple.

A few more souls arrived from the Scottish wilderness. They were a mixture of lone tourists with oversized rucksacks and older folk with flasks of hot tea. We were encouraged to put on faded life jackets before boarding the *Equinox*, a black-and-red RIB. The boat smelt strongly of diesel and I was glad when we got out to sea so that the wind could push the odour away. The captain was unsmiling and solemn, my guess being that he hadn't wanted to come out of the shed. It was a fairly short hop to wedge-shaped Handa – less than twenty minutes – but the boat travelled slowly and gently. This not only gave us more time to admire the green-topped islands dotting the Sound of Handa, but also stopped us from getting soaked by the increasingly choppy sea.

We moored at a pristine white-sand beach that was more Bahamas than Scotland. The smooth bright sand was dappled by the caress of the lapping waves and the beach was surrounded by a series of low sand dunes, making it a protected and secluded spot. We were met by two of Handa's volunteer wardens, who had rolled a makeshift wooden jetty with a converted buoy as a wheel down to the boat to allow us to alight. We were then escorted through the dunes – cue sand blowing into eyes and every other bodily crevice – to a small stone bothy. From here, I could see a few sandpipers and a lone oystercatcher picking his way through the seaweed on the shoreline and *peeping* quietly to himself. A warden told me that they'd seen two oystercatcher chicks that season but that one of them had been deftly predated by a great black-backed gull in front of her very eyes. She looked emotionally scarred by the experience. When she said that there'd been sightings of puffins too, my spirits rose. Inside, amongst the desiccated bird skulls, information boards, stuffed puffin toys and postcards, we received the usual safety talk.

We were then free to go and explore. However, in the few minutes it had taken to brief us, the weather had turned. The wind had picked up and it had started to rain. It was pretty much horizontal and I pulled on my waterproof coat before heading out. It didn't help. The rain lashed one side of my face and soaked one leg of my trousers. I pushed on alone; the others had decided to wait back in the bothy, and the flasks of tea were already out and in use. We had limited time on the island before the grumpy captain would take us home. I didn't want to waste it.

The winding path led up a slight incline away from Tràigh Shourie, through a series of wire fences, to the abandoned village. This was no more than the carcasses of a few stone buildings, with tall weeds and rampant ferns growing through them. There was some timber left too, bleached white from the sunlight and wind, and appearing like discarded bones. Though uninhabited now, in the early nineteenth century, Handa had supported a small but isolated community. This wasn't to last. The potato famine hit Handa hard and soon the island was abandoned completely.

Handa was covered in purple heather and I would have stopped to admire it if the rain hadn't been getting worse. Such downpours were seemingly not unusual, and the paths had special stone gullies to allow water to pass through them. As I continued to walk, the path became a boardwalk, protecting the sodden ground from being churned up by tourists and researchers marching up and down the island. Reeds and rushes grew in pockets alongside the path, while elsewhere large swathes of bog cotton shook in the wind, like candyfloss at a fairground.

A brave volunteer was laying new sections of wooden boardwalk along the path, despite the worsening weather. Her bright pink jacket was at odds with her green surroundings. The path continued to the horizon. I hoped she didn't have to replace it all.

I walked through some high bracken and must have surprised a snipe, as it shot off at speed with a loud *scceeeep*, taking its cartoonish beak with it. I really hadn't been expecting to see a snipe there, as they are usually found in wetlands and marshes. Thinking about it though, Handa was clearly wet enough to support its own wetlands in the centre of the island and even to have its own freshwater lochs. This was prime snipe habitat.

The snipe was not the only one to be surprised though. As I rounded a bend, I heard a loud swooshing noise over my left shoulder and saw a dark shape, blurred with speed. I instinctively ducked. I was under attack. Again, I heard the swooshing, and turned to see a great skua coming towards me, wings outstretched, body vertical, feet pointing towards me and beak gaping. It was heading straight for my face, and only turned and flew upwards at the very last minute. It was a scary, intimidating move and I was getting flashbacks to an attack in the Westfjords of Iceland again.

The great skua must have had a nest nearby, hence its behaviour, designed to protect its young. Sure, the Arctic terns on the Farnes might chastise you and peck at the back of your head, but with none of the violence or ferocity that was on display here. Or the persistence, for that matter. I tried to carry on walking, but the attack continued. The word 'skua' is one of the few that have passed into the English language from Faroese, but this was of little comfort to me. There was a pair, and they took it in turns to swoop in from a distance, directly at my head. Great skuas are mottled dark-brown in colour, with white bands on each wing and big, black feet, although the latter may just have been my perspective – they were right by my eyeballs. They even have an evil, strangled, *ha-ha-ha* laugh of a call. Great skuas are sometimes known as the pirates of the seas, and it's not hard to see why. All they're missing is the eye patch.

As boisterous as the skuas were, I was pleased to learn that Handa Skua Project commenced in 2003 to protect and monitor the islands Arctic skua and great skua populations. Both varieties of skua have bred on Handa with varying degrees of success for a number of years, although they are both currently in decline. Even pirates need a hand every now and then.

My next stop was Puffin Bay, on the north coast of the roughly circular Handa. Unfortunately, it did not live up to its name. It was a pretty enough place, a neat indent in the coast, the island's highest point before it fell off into the sea. There was a little beach below me, covered in boulders, and a small sea stack rising from the sea like a sword.

I lay on my front and shuffled forward on the cold, damp grass. From this position I could look more closely at the cliff face, which was green with ferns and rough clover and spotted yellow with kidney vetch. Small outcrops of rock bulged through, accompanied by clumps of the daisy-like sea mayweed. Directly beneath me I noticed a fall of rocks like tumbled Jenga pieces. On one of the angular juts, I saw a razorbill. And then, on another rock, in this deep, lush cliff-side meadow, a pair of orange feet. A puffin. I moved around to get a better view. It was a delightful specimen, with a glowing beak and snow-white chest. The better a puffin's diet, the brighter the colours of its beak and feet, and the more attractive it becomes to potential mates. It's the carotenoids that cause the colouration, in the same way as a ripening tomato. This puffin must have been enjoying an excellent diet. I sat and watched for a while, despite my knees sinking into the wet turf. I was so pleased to see this pristine little puffin, in his luxury home.

Between Puffin Bay and Great Stack was an intriguing sign, a knee-high post that carried a white plastic notice stating, 'Puffin Viewing Area'. It made me laugh, not only as there wasn't a puffin in sight, but because I couldn't imagine puffins abiding by such a sign anyway. I'm not sure they can read.

Great Stack was just that, a giant sea stack. It was only a few metres from the edge of the cliff but completely detached from it and surrounded by frothing sea. There was a mysterious and gloomy sea cave at its base, but Great Stack was essentially a high-rise for seabirds – four hundred metres tall and covered from top to bottom in birds. I say birds, but one species in particular had taken ownership of the stack. Guillemots blanketed the stack in vast rows along the rock's striations, making their *garrrrrrr* noise and creating a real mess that I could smell, especially when the wind blew towards me. Their paint scheme, it seemed, was white. On closer inspection, I could see that the guillemots had chicks amongst them. Soon enough, these juveniles would become jumplings, fearlessly leaping from the cliff face to join their parents – usually the father – on the sea below, bobbing around as black-and-white balls of fluff. It was not a happy story though. According to figures from the Scottish Wildlife Trust, Handa was once the largest guillemot colony in the UK but has seen a huge decline. In 1994, 113,000 guillemots nested here, compared with only 54,000 in 2016. I hope things get better for them.

At the top of the stack was a grassy plateau and a few razorbills, but no puffins. Immediately opposite the stack was a sharp V-shaped incision in the rock face, which meant that I could stand on one side of the cliff and be only a few metres from the opposite cliff. It was more sedate over there and less crowded than on the stack. Kittiwakes and razorbills occupied the occasional ledge, while fulmars rode the wind effortlessly. Towards the apex of the cliffs, on the very tops, where the thick, matted grass and thrift hung over the edges like an untidy fringe, I spotted several of Handa's puffins. They were being buffeted by the wind and clearly weren't enjoying it, but three or four stuck around long enough for me to get a good look at them. One constantly had his back to me, which could have been considered rude but was understandable given the increasing ferocity of the wind. There were 333 puffins on Handa according to the latest count. I had no idea where the other 328 were, but I was pleased to be able to say that puffins were alive and well, if not a little windswept, on Handa.

I heard a shout behind me. It was one of the wardens. I could tell by her sand-coloured fleece and red, weathered face. 'You've got to leave,' she said, her voice breathless, presumably because she'd been running around the island, herding visitors like errant sheep, and not because she was suddenly lusting after me. I took a look at my muddied knees and knew it was the former. 'The boat is going in twenty minutes.' The weather had deteriorated, and the wardens and the captain were concerned that if we didn't leave now, it might become too dangerous to collect us. I remembered the faded, twenty-year-old life jackets and decided that

I could neither risk a crossing in high seas nor face a night on Handa with no amenities but a compost toilet. I fairly legged it down those wooden walkways back to the beach and made good my escape.

My first stop on dry land was the Shorehouse Restaurant at Tarbet. This family-run restaurant prides itself on its fresh shellfish, caught within sight of the table on which it's served. It doesn't get much fresher. I had the langoustines and a cold beer. It was looking rough out there, I thought, as the rain hammered on the plate-glass window. I made myself comfortable.

14

Winner, Winner

July

Rathlin Island, County Antrim, Northern Ireland

Puffins are still revealing their secrets to us. Only in 2018, Jamie Dunning, a research student at the University of Nottingham, found something truly remarkable. Puffin beaks glow, or, more correctly, are photoluminescent under ultraviolet light. Dunning had remembered this being true of another auk, the citrus-scented crested auklet, and now discovered, when studying a (naturally) deceased puffin, that the yellow parts of the puffin's beak also lit up, and with a surprising intensity too. It's thought that this is because puffins and other birds, unlike humans, can see ultraviolet light and that a glowing beak may increase a puffin's attractiveness to potential mates. That's right – they use ultraviolet to increase sex appeal, just like a 1980s disco.

Dunning didn't stop there; he is currently collaborating with Canadian scientists to further the study. They intend to work with live puffins, so to prevent any damage to the birds' eyesight, the team has developed sunglasses. Sunglasses for puffins. I wish Dunning and his team, and the disco-ready, sunglasses-wearing, flirting puffins all the best with their studies.

Rathlin Island, my only stop in Northern Ireland, was not easy to get to. It was hopefully going to be worth it though, as I knew that puffins were frequent spring and summer tenants of Northern Ireland's largest seabird colony. I had caught an early flight to Belfast from Liverpool and then driven up to Ballycastle, past signs for Tayto Crisps and others oddly proclaiming 'Winner, Winner, Chicken Dinner', whatever that meant. I decided that I would definitely use the phrase more often, in any case.

From Belfast it was a fifty-five-mile journey through Ballymena and along the strip of dual carriageway that passed for the M2 to Ballycastle on the north coast, not far from the famous Giant's Causeway, whose hexagonal basalt columns mirror those I'd seen in Staffa. Ballycastle seemed a pleasant sort of a place, sitting on the Antrim coast in the midst of deep green fields and sandy beaches. I quickly sourced a bacon sandwich and a steaming coffee in the town – sorry, the *Sunday Times* Best Place to Live in Northern Ireland 2016 – before heading down to the ferry port. This was next to the harbour, but the signs were unhelpfully hidden, resulting in me taking a brief tour of Ballycastle's finest parking spots before I figured it out. I was not helped by the car radio picking up Radio Scotland

and giving me up-to-date traffic reports from Glasgow city centre. There was a newish building, all glass and painted steel, as a booking office, but my ferry was about to depart, so I had to run to the waiting craft, which was already laden with vehicles and people bound for Rathlin.

In contrast to Ballycastle and the ticket office, the ferry was somewhat dilapidated. It was a small vessel, with white paint between the rust patches, a green metal floor and just one deck, open to the weather. I joined the other passengers on a row of loose plastic chairs that seemed to have been borrowed from a village hall. Most of the deck was taken up by three vehicles, all anchored down to rings in the floor but bouncing about on their suspension – a knackered Volkswagen Golf that had recently been used to plough fields, and a cleaning supplies van that was, ironically, towing a cattle truck liberally covered in cow shit. My fellow passengers, mainly German tourists, were wrinkling their noses, either at the smell or in the hope that the swell wouldn't get any rougher. The latter concern may have been exacerbated by the casual attitude to on-board safety, including the cupboard marked 'Man Overboard Recovery Pack in This Locker' and the life-ring stuck under a pile of metal and other debris in the corner. There was not a life vest in sight and the crew were up on the bridge, all in fluorescent orange jackets, sharing freshly received iPhone jokes. The six-mile crossing took about twenty-five minutes, under a dark grey sky and chased by a sharp wind.

Rathlin appeared low-slung in the sea, with the Rue Point lighthouse like a dropped Christmas liquorice. Rathlin has always been problematic for ships or, rather, for mariners. I've learnt that the number of lighthouses on an island usually corresponds with the number of hazards and wrecks around its coast. There are two other lighthouses on the island – West Light, which I was heading for, and East Lighthouse.

The ferry abandoned us at a small village, whose pretty harbour was festooned with lobster pots and enclosed by a stern stone wall. Coloured bunting rattled in the wind, around several expensive-looking boats. I assumed all the working boats were out to sea. There was a car in the harbour car park that was clearly less of a conveyance and more of a farmer's shed. It had four flat tyres and clumps of grass growing from the wheel arches. Inside, the seats were barely visible beneath the tools and wire, twine and fertiliser bags. It had once seemingly been a silver Isuzu, but I couldn't be sure.

The village was home to most of Rathlin's 140 residents and seemed very pleasant, with several whitewashed buildings forming a crescent around the bay, and the church standing proud. There was a post office and shop, a couple of artist's studios, a café, and a pub called McCuaig's. This turned out to be dingy and dour and, as much as I fancied a pint of Guinness, I couldn't stand the low-rent Wetherspoon's feel of the place or the smell of stale beer, so I settled for a packet of subliminally advertised Taytos from the shop instead, and a creamy farmhouse ice cream served from a converted horsebox by a kid just big enough to reach the freezer. No sign of a chicken dinner.

There were several puffin themes going on around the village, including a large, oddly one-footed puffin statue by the harbour (next to the Aquaholics boat-trip sign), plastic puffins peeping from tubs of marigolds, the lesser-known lawn puffins, and puffin ornaments of every type in the shops. It was also Bert's Puffin Bus that transported visitors to the RSPB centre at West Light. Bert, I assumed, was the older chap who was counting passengers onto the small minibus – and failing, as one couple had to share a seat. They didn't appear to have done this in a while and the woman looked particularly displeased.

The Puffin Bus had seen better days and was seemingly held together with rust and spiders' webs. A female driver took the wheel, complete with over-practised and dated tourist commentary – 'the motorway ends here', 'Fred Flintstone used to live in that cave' and so on. She was friendly enough though, even if she had to shout to be heard over the dilapidated engine. And besides, those four and a half miles would have taken ages to walk – the route twisted its way across the L-shaped island, at times travelling down lanes lined pink with fuchsias, at other times taking us across windswept moors. The island is seven miles long and never more than two miles across. It felt much bigger though, and had the terrain of its Scottish neighbours.

We passed Rathlin's tiny school, with only ten pupils, and fields glowing green with cowslips and ferns. It was hot and humid, and the wind had stopped. I had the uneasy feeling of being in a moving greenhouse and mopped at my attractively glistening brow. We stopped briefly to admire the view at Knockans, back over the sea to cute Ballycastle, and then continued on to the stunted woodland at Kinramer. I was also keeping my eyes open for one of Rathlin's more elusive characters – the golden hare. A close relative of the Irish hare, this beauty is golden, nearly ginger in colour, and some individuals have bright blue eyes rather than the sombre browns of their cousins. Although not native to Rathlin, the golden hare is now only found on the island and is thought to be the result of a rare genetic mutation. Puffins can display the results of genetic mutations too. Leucistic puffins are extremely rare, but show as nearly all white due to a lack of colour pigmentation. Individuals have been spotted on Skomer and in the Isles of Scilly in recent years, and have made the news as a result. Catching a glimpse of one of Rathlin's supermodels would be a rare treat though.

We arrived at West Light and were given a couple of options for the return journey. I vowed not to miss the bus as the walk back would be torturous in the humidity and would potentially mean missing the return ferry.

The RSPB centre at Northern Ireland's largest seabird colony was a white, glass-fronted building perched on the edge of the island. It was busy with a lot more blue-uniformed RSPB staff than I had expected, which was a little overbearing, so I skipped the information boards and the attempts to get me to join the RSPB – I was already paid up – and headed straight through.

On the other side, there was a clear view of the coast, a jagged series of peaks rising from a serene sea. Even the sea appeared languid in the late July heat. As I descended the several storeys of steep concrete steps to the lighthouse and bird colonies, I could tell I was getting close; not only could I hear the birds, which made me smile from ear to ear, but I could smell them too. The deep, pungent odour hit me like a fist to the nose.

West Light was essentially built upside down, and the ninety steps that zigged and zagged around and down the concrete slope were part of it. The lighthouse clings to the cliff limpet-style, rather than towering above it, and this allows the light to be at the optimum height for warning passing ships. The lighthouse took seven years to complete and included having to first build a pier, a road and an inclined railway to transport materials. West Light's now disused foghorn, known as the Rathlin Bull, was supposedly so loud it could be heard back in Ballycastle and beyond.

At the base of the lighthouse was a platform area given over to birdwatching. This too was patrolled by an army of RSPB volunteers, who pounced with great enthusiasm on each person wheezing down the stairs. I later found out that they were training new volunteers, hence the number of them around. I suppose they had to learn somewhere, and I was anyway glad of their help. They pointed out the razorbills and guillemots on the lower cliffs. The guillemots were, as usual, the noisier and more uncouth residents, living in close proximity to each other and frequently quarrelling with their neighbours.

At the bottom of the slope was a slightly less steep area, worn smooth by the constant paddling of orange feet and temporarily void of vegetation save for a few scraggly clumps of grass. This was puffin territory. Safe from land-based predators by virtue of its location, it was also graced with sufficient soil to dig some good old burrows and was near enough to the sea for food and yet not so close as to be at risk of flooding. Twenty to thirty birds had gathered there. Some were still coming in from the sea with their moustaches of fish, their white chests reminiscent of the tucked-in napkins of Italian gangsters at a fine restaurant. Others were taking things more leisurely, standing defiant outside their burrows, facing out to sea. There were a few puffin quarrels, but we were coming to the end of the season now and most disputes seemed to have been settled; it had probably been a lot busier there a couple of weeks earlier.

Their orange beaks and feet made the puffins stand out against their sombre backdrop, but I wanted a closer look. Lacking a head for heights and having no equipment save for a packet of Taytos and a battered rucksack, I dropped any ideas of abseiling down the sheer rock face and, instead, borrowed a telescope from a friendly RSPB volunteer. I am a member, I should stress – lending out telescopes is surely the least the RSPB can do for their acrophobic members.

The telescope gave me an excellent view of the puffins at the base of the slope. It wasn't the same as at Skomer or on the Farnes, where the little critters had run around my feet, but I'm always so glad to see them. I studied one puffin which was absolutely pristine. His white chest was plump and full, and the white around his eyes was uninterrupted. I surmised that the sheer whiteness of a puffin's plumage might one day be adopted by Mexican drug cartels to describe their purest cocaine. 'An ounce of puffin' would become an all-too-familiar phrase. This puffin had the classic wonderfully-coloured beak, a real prize specimen, but I was transfixed by his feet and legs, which were fantastically orange. Puffins' legs are much further apart then you'd think. This helps them stand steadily on precarious places such as clifftops, and also in the water, where they become super-effective rudders. He was clearly a very healthy bird. I've seen puffins that have been rescued or kept in captivity, and their yellow, dull feet and lacklustre, cracked beaks always give them away. This Rathlin puffin was thriving, as were all the birds I could see. He paraded around, and stood tall and alert.

I couldn't resist using my iPhone to try to take a photo of the puffins through the telescope. This was harder than it sounds – no, really – and I was grateful when a twenty-something volunteer with several prominent piercings offered to give me a hand. 'You have to hold the scope with one hand,' she said, 'and then angle the phone with the other.' I tried as hard as I could but gave up with the scope, much to her amusement. We chatted for a while, and then she divulged her secret. 'Come with me,' she said. I followed her to the end of the viewing platform. 'You won't need the scope now.'

She pointed out a beautiful fulmar chick. It was only about three metres away, a gorgeous white powder puff. She'd been watching the chick since it had hatched three weeks earlier; at first, the parents had been attentive, but now they'd started to leave it for extended periods. It looked well fed and content. It had sharp pinpoints of black for eyes and a stubby but dangerous-looking beak. I was wary of any spitting. Its nursery was on a narrow ledge, out of the wind and out of view from other predators.

I returned the favour by telling Louise that, in Iceland, fulmars are known as the 'elephant bird'. She laughed. 'No, it's true.' I explained that the Icelandic word for fulmar is easily confused with the word for elephant – *fill*. It was a pretty useless fact, I have to admit, but if you go to Rathlin and an RSPB volunteer tells you about the majestic Icelandic elephant bird, at least you'll know where that came from.

I stayed and observed the fulmar for a little while. Like any good toddler, it spent its time either squeaking for its parents or dozing. Winner, winner, chicken dinner.

15

The King of Lundy and His Half-Puffins

July

Lundy, Devon, England

Lundy is a lonely island in the middle of the Bristol Channel, somewhere between Devon to the south and Wales to the north. It actually takes its name from the Old Norse for puffin – *lund* – and island, *ey*, as puffins were once ubiquitous there. I was going to see if the island could still legitimately live up to its name. These days, Lundy has a human population of around twenty-eight, and its own pub, but most importantly it also used to have its own currency of 'puffins' and 'half-puffins'.

During the winter, Lundy is accessible only via a short but hairy helicopter ride but, during the summer, it is visited by MS *Oldenburg* four times a week. The ship sails alternately from Ilfracombe and Bideford, and the twenty-four-mile journey takes about two hours. It is not the fastest ocean-going craft, but it's sturdy and dependable, as I saw when I arrived at Ilfracombe while it was being loaded up. I was glad to note that a dray from St Austell Brewery was ensuring that eight barrels of beer would be travelling with us.

Ilfracombe is a pretty seaside town, enhanced by being on several levels and perched on a rugged, beautiful stretch of the Devon coast. Its rows of Victorian houses nestle neatly into a lush green backdrop, while down at the harbour, overlooked by the charming St Nicholas' Chapel, there's a surprise waiting. 'Verity' is a statue created by Damien Hirst, he of halved animals in formaldehyde. It is certainly striking. Marked by herring gulls and turned blue by salt and rain, it depicts a naked pregnant woman thrusting a sword into the sky. On one side. On the other, the woman's skin appears to have been removed to display her skeletal and sinewy interior, and, alarmingly, the unborn child. Towering over the harbour, it is quite a thing, especially when set against the quiet, gentle, seaside sensibilities of Ilfracombe. It's brave, but I liked it, particularly when viewed from the sea, which is where it has the biggest impact.

The harbour was filled with a huge queue of people waiting to board. Those staying on the island for a few days had bags and suitcases, while day-trippers like me had very little. We watched the ship's crane hoist boxes of luggage aboard and wondered why we'd brought waterproofs and not sunscreen. The early-morning mist had burnt off to reveal that most un-English of things, a perfectly blue summer sky. It promised to be a hot, sunny day.

We were eventually permitted to board by a crew wearing what appeared to be fancy-dress sailor costumes. They had epaulettes, clip-on ties, and shirts that were in need of a good iron. We each exchanged our outbound ticket for another ticket, no explanation given. The ship was a fair size, but there was little room to spare, so I was grateful to grab a seat on a lifeboat box on the upper deck. The ship had cream funnels, but the masts and other areas had been painted rust brown. It was either a strange or clever choice of colour.

After the safety briefing – put on warm clothing before life vests, keep your head in while leaving port – we were off. The *Oldenburg* was soon at sea and belching the smell of diesel and bacon sandwiches from below. I would later visit the bar area to find that it was stuck in the 1980s, with no apparent updating of decor or foodstuffs. You could still buy Opal Fruits and Marathon bars at the shop. I decided to stick with the scenic but rugged views of Devon, seen from my perch next to the door to the bridge, which was open. I could hear the captain gossiping about the crew, and, worryingly, the sound of something being repeatedly struck with a hammer.

There was sadness here too. On an early August day a few years back, when the *Oldenburg* was in dock at Ilfracombe, a member of the crew tragically lost his life when he became trapped between the ship and the harbour wall. It was difficult to imagine how this happened, or how the crew continued on, greeting and smiling at new passengers every day.

As we passed Woolacombe beach, there were surprisingly few houses on the shore, and only the occasional

184

white boxes at caravan sites; otherwise, it was just greenery and a lighthouse on each point. I gazed out at the sea, where the sunlight was hitting the gentle waves and making them shimmer like spilt glitter. A group of old men were already on their third pint, but I was happy with my sunshine and sea spray.

About an hour into the trip, when I had become pretty much hypnotised by the rocking of the ship, and the excitement of getting greasy sandwiches had waned for most families, there was a shout from the bow. The captain came scuttling through the bridge door and pointed out to sea. 'Dolphins!' he bellowed. I looked out to see a pod of four or five common dolphins at the bow of the ship. They were jumping in and out of the water, in pairs, and each time they did, I could see both their white underbellies and their slightly bemused expressions. They put on a show for a few minutes, crossing under the boat and causing further gasps of excitement each time they re-emerged in the swirling, frothy surf. It was a pleasure to see them. It certainly showed that the ocean was clean and productive enough to support these wonderful creatures. It is always much more impressive to see them in the wild, in their natural habitat rather than in a Floridian tank, and almost twice as impressive when seen off the cold shores of our own island. It's just so unexpected.

Lundy was much longer and higher than I'd imagined. It was hard to miss, rearing abruptly out of the sea like a pile of soil dropped by a careless gardener. The sun was beating down, but I had my coat on as the breeze from the sea was chilling me, especially when I leant out of my sheltered spot to get a better view of the island. I could see a white lighthouse and a church. In the sea, I spotted the odd guillemot and, above us, a couple of herring gulls, but there was a definite lack of birdlife. I was worried that perhaps I wouldn't see anything at all, let alone puffins.

We landed on the south end of the island, adjacent to Rat Island. Beyond Rat Island was Devil's Kitchen, and no one wanted to end up there – although the two place names did have me singing a UB40 song in my head. Rat Island is separated from the rest of Lundy by a short shale beach, leaving a narrow gap through which I could see the royal-blue ocean. Above us was South Lighthouse, a modern lighthouse in sparkling white, like a wedding cake. We all piled off and trudged up the wide jetty like a scene from the Ark. The harbourside was surprisingly well developed, with a new building clad in sleek larch, and various tractors and trailers, as well as Land Rovers to ferry elderly people and luggage up to the village area. I wouldn't have minded the Land Rover myself, as the path up the hillside from Victoria Beach was steeper and longer than I'd expected. I passed a lonely cormorant on a rock; the higher I got, the more the view improved, and the more I panted.

In 1924, after centuries of piracy and worse, the island was bought by the eccentric Martin Coles Harman. Harman had fallen in love with the island as a teenager and was evidently quite a character. He insisted that Lundy was its own state; he even set up his own customs point and introduced his own puffin-themed currency. The coins eventually saw Harman prosecuted and convicted. As I walked around the island, I kept an eye out for any coins on the ground. I would have loved to have found a genuine puffin coin on Lundy. I didn't though, and ended up buying a half-puffin from eBay once I was back on the mainland. It's not often you get to bid on a half-puffin and win. I'm intrigued by the little bronze coin, with its puffin head still clearly visible. It seems like Harman shared my obsession with puffins.

Harman also introduced puffin-adorned Lundy stamps to cover the cost of postage back to Devon – rather cutely known as puffinage. There was a smart red post box by the shop in the village, and you could buy stamps easily at the shop. The mail used to be carried by sea or air but, these days, the *Oldenburg* takes both the postcards and the tourists who have just paid extra to post them on the same ship. Oh, well.

At Marisco Castle, an impressive stone structure built by Henry III in an attempt to make his mark on the island, I followed the path west away from the crowds trying to find both the pub and the shop, and away from the Land Rover that was bombing back and forth, creating vast plumes of dust in the air. No one else, I noticed, was heading west. This suited me just fine. I walked the south edge of the island, past the wonderfully named Rattles and Seals' Hole to Devil's Limekiln and Shutter Point. Here, the island had shattered into a sharp point, and jagged pieces lay in disarray at the bottom of the cliff. I continued walking up the west side, known helpfully as West Sideland, using Old Light lighthouse to guide me.

There was more than a slight problem with Old Light. Not fully considering the islands natural height, Old Light was constructed way too tall and was more often than not concealed by the weather, rendering it completely useless as a lighthouse. Old Light remains a handsome, well-built place and is now hired out as a holiday home. The light has been replaced by two canvas deckchairs, which must give wondrous views across the island – or the inside of a raincloud.

I continued walking up the path, heading north. At Ackland's Moor, I was pleased to spot the plentiful yellow flowers and thick green leaves of the Lundy cabbage, only found on Lundy itself. Past Quarter Wall, at the craggy rocks of the ominous-sounding Dead Cow Point, I could see across the boggy Punchbowl Valley to my destination, Jenny's Cove. A wheatear closely followed my progress, hopping along only a few metres to my left, cheerfully chattering away to itself or me, it wasn't clear which. I also encountered a few of the island's Soay sheep who were too engrossed in the grass to watch me pass. Named after Soay, an island off the far-flung but also puffin-adorned St Kilda, they too were introduced to Lundy by our old friend, Martin Harman. The Soays are tough as old boots and can graze and survive in the harshest of environments.

Jenny's Cove was the place on Lundy to see puffins. I knew this because of my own research, and because there were big signs on the island stating that this was the 'last chance to see puffins'. I assumed they meant that season, not ever. There was a small gathering of people just around from the tall granite stacks oddly named the Cheeses. These didn't look much like cheese to me, more like custard creams as stacked by a toddler. I was evidently in need of lunch. Jenny's Cove was named for a ship, wrecked there and said to have been carrying ivory and gold. The story goes that it wasn't until decades later that a message in a bottle washed up at Babbacombe on the Devon coast advising the finder to search Lundy for the lost valuables. The ivory was subsequently found, but the gold never was. I chuckled to myself at the thought of the puffins using the gold to bling themselves up. It really had to be lunchtime soon. The lack of food was making me delirious. Puffins, I thought, then pub.

The gathering comprised of a few day-trippers and holidaymakers. The Landmark Trust, which manages the island, had kindly set up a telescope there, and I used it to watch the colony on the south side of the cove, just before the Devil's Chimney sea stack. In 2005, it was estimated that there were only two or three breeding pairs of puffins here, numbers having plummeted due to black rats, which see eggs and chicks from burrows as an easy meal. Since then, however, Lundy has had a rat-eradication programme run by the Seabird Recovery Project. It became rat-free in 2006 and the seabird population has bounced back, with the RSPB reporting in 2019 that there were now 5,500 pairs of Manx shearwaters on Lundy, along with 375 puffins.

I was as pleased as Punch to see several puffins out and about at Jenny's Cove. I sat and watched for a while, despite my rumbling stomach. There were some younger birds there, darker around the eyes and slightly slimmer, their white bellies not yet fully rounded with what are actually powerful chest muscles. These birds were only three to four years old and hadn't bred that season, but were prospecting for burrows, and mates, for next year. I hoped they would come back to breed; despite the success of the Seabird Recovery Project, the RSPB reports that Lundy's current puffin population is only one tenth of what it was in 1939.

I managed to find a copy of Richard Perry's *Lundy - Isle of Puffins* in a second-hand bookshop. First published in 1940, he writes fondly of the birdlife on Lundy and especially the puffins although, ominously, Perry pays thanks to his 'overlord' Martin Coles Harman. It struck me that his writing was from the 1939 puffin season – exactly the same year the RSPB had chosen to draw comparison to. Perry spent five months on Lundy studying the breeding season of the puffin, often endearingly describing them as 'dancers'. Perry writes, 'On the opening day of July the puffins entered upon the last stage of their occupation of the island in 1939. At eight o'clock that morning, with a heavy sea running, there was a full house of both puffins and razorbills on Puffin Slope and a wonderful flighting of a thousand birds slipping in sideways from the sea en masse after streaming over the boulders.' He goes on, 'It is sad that the pen can record only prosaic, colourless memories of the thousands of red, white and black little figures dancing on the olive-yellow lichens of the grey boulders and the bright green cushions of the thrift. After watching guillemots and razorbills intensively for some days, the brilliant vermillion paddles of those puffins still in full health hit the eye with an unusual vividness: one seemed to be seeing bright-coloured birds for the first time.' I'm not sure I could put it much better myself.

As I sat there admiring Lundy's impressive little birds, a huge grin appeared on my face. There was something so special about seeing them thriving on their eponymous island, as I'm sure Perry would agree, although he'd be surprised at their depleted numbers. The puffins were a sociable bunch, appearing to chat to each other in that low, cartoonish growl as they slowly moved around. The older birds were alert and obviously on the lookout for predators, such as gulls. Looking at the young puffins, I thought of them as a bunch of teenagers in a pub beer garden, trying to strut their stuff while Mum and Dad had a drink and a packet of crisps inside.

My thoughts had clearly returned to food and pubs, so I handed the telescope to the next in line and set off down the Quarter Wall, enjoying the sunshine and the thought of a pint, something to eat and then maybe even a snooze before it was time to board the boat. The path headed south, and it wasn't long before I saw the tiny village before me, a picture-postcard cluster of buildings hewn out of light-grey Lundy granite. On the high street – if you could call it that – I popped into the shop that sold everything from calabrese to Cornettos. I settled for a Lundy stamp bearing a handsome Lundy puffin. There was no way this little beauty was getting posted anywhere.

The Marisco Tavern had a nice green patch of grass immediately outside its front door, I noticed, for potential snoozing later. Inside, it was welcoming enough. The walls were festooned with all manner of Lundy artefacts, mainly nautical, including lifebelts from vessels long gone, bearing legends such as 'Greenland Trading Post', oars and tens of coats of arms. There was a tall cupboard full of board games and another full of books. Clearly, the Marisco was somewhere people could get marooned and not really mind. It was a reminder that the trappings of modern life, and modern pubs, had no place here. There were no mobiles to stare into, no gaudy fruit machines and no Sky News. This was no bad thing. The Marisco Tavern is known as the pub that never shuts, it having to serve as the island's base. It's also the only building to have twenty-four-hour power and lighting when generators shut down everywhere else. I could see why some visitors to Lundy never get further than the Marisco's bar.

I was served by a barmaid with a pronounced Bristol accent and ordered a pint of St Austell Old Light Ale, specially brewed for the pub and delivered with me on the *Oldenburg*. I couldn't resist ordering one of the homemade 'Cornish' pasties containing Lundy lamb either. I made my way to a patch of grass not too far from the pub and sat quietly in the lush pasture. I could see the Landing Bay from there, and I was sorry to think that I would soon be leaving this special little place. I quite understood why Martin Coles Harman so loved it, and I was pleased that Lundy was still thriving, thanks to his input. I supped on my beer and tucked into my pasty. It was fantastic, by the way – really tasty lamb, potatoes and swede, with plenty of pepper. It doesn't get much better than this, I thought. The sun warmed my arms and I started to relax. Mind you, that Soay sheep was looking at me funny. I hurriedly gobbled up my pasty.

16

A Sky Full of Puffins

July

Shiant Isles, Outer Hebrides, Scotland

To be brutally honest, I'd never heard of these little islands
on the edge of nowhere until the BBC documentary series
The Last Seabird Summer?, in which the islands' current,
ever so slightly eccentric owner, Adam Nicolson, spends a
season tracking the Shiants' seabirds. Because I'm British,
I've always watched the BBC. From childhood, it's been a
staple of my life – in fact, many school friends were forced
to watch only the BBC, never the bawdy ITV – meaning
that it's often fed my imagination and provided inspiration.
Never more so than with *The Last Seabird Summer?* I
consumed the two one-hour documentaries whole, and
then watched them again immediately. As an introduction
to the Shiants, they were awesome. As a story of seabird

decline and change, they were deeply unsettling.

Adam has also written a book on the Shiants – the beguiling *Sea Room* – look up a copy if you get the chance. He was given the Shiants by his father, who had bought them with an inheritance after spotting an advert for the islands in the *Daily Telegraph*. Adam goes on to make the point that while he is the legal landowner, the Shiants don't belong to anyone; he is merely their keeper. They've been owned and ruled by many over the years, but the islands themselves don't change, don't care. Adam is at pains to stress that he is their custodian, looking after them and making sure that their rare beauty is shared as widely as possible, while being kept as safe as can be. It sounded quite a task, and the more I read about the Shiants, the more I wanted to see them.

In *The Last Seabird Summer?*, Adam pays a visit to the puffin capital of the world, Iceland. Just like I have – many times. Here, he sees both declining and thriving puffin populations. In the Westman Islands, he visits a family who hunt puffins with a large sky net, using a very similar approach to that employed by researchers on the Shiants. As I had seen, the Westman Islands have seen whole colonies of puffins disappear from the slopes on their islands, and the family is no longer able to catch puffins where they were previously plentiful. In the far north of Iceland, he travels to Grímsey, an island bisected by the Arctic Circle, where the puffin population is booming and local planes have to pass the island twice just to clear the seabirds from the runway prior to landing. Again, Adam meets up with a local family who catch and eat puffin. It's grisly viewing at times, especially when the breast meat is being removed by hand, for a barbecue later. 'Is this a traditional Icelandic recipe?' Adam asks as the breasts are rubbed in a rich sauce before being taken to a red-hot barbecue.

'No, it's just American hot sauce,' comes the reply.

I had my first glimpse of the Shiants as I flew from Aberdeen to Stornoway on the Isle of Lewis in the Outer Hebrides. They appeared as three dark-green blobs of moss floating in the strip of sea known ominously as the Minch. There are three main islands, House Island (Eilean an Tighe in Gaelic), Rough Island (Garbh Eilean), Mary Island (Eilean Mhuire) and a smudge of smaller skerries and stacks to the west. The Shiants are only five miles from Lewis, but they may as well be five hundred miles. In the other direction, it is twelve miles to Skye. It was a sunny, blue-skied day up above me, but I noticed that the Shiants were swirled with cloud. It reminded me of a remote, undiscovered volcanic island, something like Jurassic Park but with fewer prosthetics. I wouldn't have been surprised to see a pterodactyl surfacing from the mist, but no such luck. I gave an involuntary shudder and then munched down on the Tunnock's caramel bar and strong tea that they still give out on these island flights.

The Shiants are not easy to get to. There are no scheduled boats, mainly because the islands are uninhabited and are a Special Protection Area. There are some boat trips from Stornoway, some of which promise the Shiants but are not permitted to land. Others are far too expensive. I did some asking around Stornoway, and the name of one man kept coming back – 'Sure, Joe'll take you,' was the oft-repeated reply, followed by a hastily scribbled mobile number on a scrap of paper. I gave it a call. Joe was going to the Shiants tomorrow to pick up a film crew. Yes, I could tag along. Did I have my own waterproofs or should he bring an extra set?

He told me to meet him at the petrol station owned by his family at the edge of town early the next morning. I was thrilled that Joe had agreed to take me along with him. I celebrated by going back to the hotel and indulging in a roast-beef carvery and a couple of pints of Islander from the town's Hebridean Brewery Company. The brewery also does a beer called Berserker, which seemed apt, but the hotel bar didn't stretch to that. I could barely sleep that night, what with the anticipation of finally landing on the Shiants, images from the BBC in my head, and a stomach full of beef and beer. It is estimated that over sixty thousand puffins breed on the Shiants, nearly 10 per cent of the UK puffin population. Every one of them filled my dreams that night.

After an early start and a full Scottish breakfast, including fried potato cakes – one of those delicacies that only ever taste good away from home – I drove out to the Engebret petrol station on the far reaches of Stornoway. It was a cloudy, overcast day, and I hoped that it wouldn't rain. I was met by a friendly attendant who told me that Joe wasn't in yet, but she could make me a strong cup of tea to keep me going. I took her up on the offer and had a wander around the shop, which seemed to sell exactly everything anyone would need while living in the Hebrides, in particular anyone who was up for a bit of hunting, fishing or shooting. Rifles stood guard in tall glass cases, while fishing rods reached over crisps and biscuits. Knives and ammunition were safely locked away. The section that I was drawn to was the selection of salmon and trout flies. These colourful and clever creations are used by fly fishermen to imitate or excite fish into taking a nibble, whereupon the gullible fish is then hauled in on the hook hidden inside. The flies had wonderfully exotic names such as Minister's Dog, Ally's Shrimp, Parachute Adams and Woolly Bugger. Some were dowdy and unexciting, while others had neon stripes and bulging eyes.

It was while I was looking at the flies, and reliving my own fishing exploits, that Joe approached me. He was a mountain of a man, with a strong, bone-crushing handshake to match. He had a ready smile and apologised for being late while simultaneously summoning me to follow him through the stock room, which looked like Del Boy's front room, with stock of all sorts falling out of boxes and from wall-to-wall racking. He collected a bundle of waterproofs and explained that I wouldn't be the only passenger on the trip out to the Shiants that morning. 'There's a Belgian girl coming too,' he said. 'I promised her boss that she could come for a trip.' Joe and I chatted some more, and he told me about his Norwegian ancestry, the trips to his homeland and his continuing efforts to enjoy the infamous lutefisk served with butter and bacon. We swapped tales of grisly meals for a while, while he threw back gulps of scalding hot coffee. I liked Joe.

I followed him down to the harbour, which was full of bobbing white boats. I carried an outboard motor down to the RIB for him, while he went to collect my twenty-something shipmate, Victoria, and kit her out with waterproofs. I chatted to Victoria for a while, but it was difficult, in part because the wind was rising and kept both whipping her words away and wrapping her hair around her face, but also because she was trying to explain what she was doing in Stornoway. It was something to do with monitoring the growth of rhododendron plants at the handsome Lews Castle across the harbour, which she kept gesturing to. I couldn't grasp why she'd come from Belgium to do this, or why rhododendrons would require a trip to the Shiants. I decided not to worry about it and instead hopped aboard and straddled one of the seats like a small horse.

The RIB was red and black with silver seats, and I was reminded of children's fairground rocket rides. I guess it was just the colour scheme. Victoria sat next to me. There wasn't much of a safety briefing, but I pulled on a life jacket, feeling like an expert now. Joe gave us a brief tour-guide bit and, ominously, radioed our departure to the coastguard. We headed out to sea. 'It's fine here,' said Joe cheerfully, 'but when we get past the headland, it will get a bit choppy, and the weather's coming in.' I tightened my life jacket against my chest, but it did little to help.

We headed south, hugging the coastline as Joe skippered from the front, the boat hopping over the waves like a skimming stone. We passed small coves, wide loch mouths and tiny islands sprouting stunted pines. To the east, the Minch looked calm, but the clouds were building, and I could see concern on Joe's face. Near to the small isle of Eilean Liubhaird, we started to pull away from Lewis. I saw the Shiants in the far distance, rising out of the sea. It was difficult to focus though, as the sea was getting choppy, just as Joe had predicted. Each time the RIB hit the water, I got a shower of the Minch – freezing cold salt water. My cheeks began to burn, and my eyes streamed in the strong wind. This was not travelling in comfort or luxury, and I silently urged Joe to get us there safely and quickly.

The RIB skipped over the waves, spending more time in the air than the water. Conversation was now impossible, and I saw that Victoria had her eyes closed and was holding on for dear life. The Shiants did not seem to be getting any closer, and I wondered who had ventured out here first, making this treacherous journey in small wooden boats. I knew that this area of the Minch was deadly, with swirling undercurrents and an unpredictable nature. It was not hard to imagine sailors' lives being lost.

We approached Rough Island first. Joe was keen to get ashore to collect the film crew, but he wanted to show Victoria and me a bit of the island too. He started by taking us to a vertical basalt cliff face. The basalt was as flat and dark as a gravestone, rising 120 metres into the solemn sky. The rocks were similar in formation to the angular Giant's Causeway, but there weren't the coachloads of tourists and no one had yet tried to sell me a Shiants eraser or T-shirt.

Joe skilfully manoeuvred the boat so that we were directly beneath the cliffs, which made looking upwards a dizzying experience. From here I could see that the strips of granite were mottled with luminous green growths of seaweed, and, further up, the giveaway white streaks of seabirds. Guillemots and razorbills in the thousands were using the cliff as their home, and the sea around them was being peppered with guano. It seemed like less of a welcome, more of a warning, as did the pungent smell created by this temporary city of birds.

Joe moved us along the cliff face and into a small cave. Here, the water was a shimmering turquoise, and Joe wanted to show us something special. He waited until a guillemot dived from a nearby shelf and then pointed. The water here was so clear, we could easily see the bird swimming beneath the surface, turning almost into a fish, streamlined and with the last of the air being squeezed from its feathers as traces of bubbles. It was utterly mesmerising watching these birds transform from noisy, ungainly beings crammed on a shelf, fighting for space, into these super-slick, underwater lightning bolts.

We stayed for a while, watching in silence. Time after time, the surface was slashed like a knife through butter, guillemots turning into fish as they dived deep for their lunch. Guillemots can dive down to 180 metres, three times as far as a puffin. The Brünnich's guillemot can dive as deep as 192 metres. For such small birds, this ability to dive and hunt at such depths is truly remarkable; it's well beyond the diving capabilities of humans, and that includes the ones all knitted out in the very best diving gear. Seeing it first-hand, less than a metre away from the side of the RIB, was something else again.

Joe shouted and pointed upwards, his words echoing off the walls. He'd spotted something worth shouting about – a small kittiwake chick. It was a fluff ball, to be honest, just peeping out of the rudimentary nest prepared by its parents. Joe told me that he'd seen up to three kittiwakes per nest, but not recently. He was pleased to see this one though, as kittiwakes are a barometer species for seabirds – if seabirds are struggling in an area, kittiwakes are always the first ones to go. This is often due to a lack of sand eels, something that affects puffins too. I often think of kittiwakes as delicate gulls, but they are actually pretty hardy. I love that kittiwakes are proper *sea* gulls and spend their winters out at sea. The chick was sitting still, looking at us looking at him. I had no idea what he was thinking. The kittiwakes above were being wound up by a fulmar, which was busily swooping around the cliffs, causing them to cry out in turn. It was like a Mexican wave of sound.

Joe chugged us slowly around so that we were heading towards the narrow isthmus linking House Island and Rough Island. To our right was Toll a' Roimh, a wonderful sea arch, reflected in the sea below. Bare stone framed the arch itself and I could see the grass rippling in the wind above it. The broken stumps of other arches, now claimed by the Minch, poked from the water. My attention, though, was neither on the sea arch nor even on the sea. It was on the sky. The sky was full of seabirds, and mostly puffins at that. It would be an exaggeration to say that it was dark because of the birds, but not much of one. There was a distinct wheeling of the birds, and I could almost hear the flapping of thousands and thousands of tiny wings. They were not even that high; some of them careered past my head, so I struggled to turn quickly enough to catch a glimpse of orange feet, white underbelly or that holly-berry red of their beaks. I had never seen so many birds in one place, let alone so many puffins. They swirled and circled above us. It was the most captivating thing.

It should have been possible to see the sky through the narrow gap between House Island and Rough Island, but all I could see was a twisting, twirling carousel of puffins, blocking my view in the most fantastic way. It was a joy to see so many. The Shiants are clearly a puffin stronghold, and rather than spending my time spotting the odd puffin here or there, as I had done in some places, a motorway of puffins was now obscuring the sky. I continued to watch them. The returning puffins had great beaks full of shimmering, silvery sand eels, and the outgoing puffins were quicker off the mark. The sea, too, was full of puffins. Dark little shapes bobbed happily on the sea, oblivious to the choppy conditions and clearly used to much, much worse. Puffins have to flap their wings up to four hundred times a minute to stay airborne and can fly up to fifty-five miles per hour, which is impressive but also rather arduous; they are much better at swimming. The aerial puffins *ffffft*-ed past my head, a blur of black, white, black, white, unless I studied an individual hard enough to pick out the colour. It was exhilarating. My heart was hammering away in my chest. This was wildlife up close and in my face. The puffins were raising a middle finger to extinction here in the Shiants.

With the engine killed and silent, we drifted towards the shore. This was the fiercely steep slope of Carnach Mhòr on Rough Island. It was as green as a cricket pitch at the top of the slope – if cricket pitches came in such acute angles. There was then a band of scree, no doubt caused by the erosive, violent winter weather and, nearer the shore, a confusing mix of jagged pieces of basalt, jumbled like broken biscuits. The whole slope was covered in seabirds –

guillemot families living next door to razorbills and shags. Puffins were plentiful and seemed to have made use of both the grassy slope, where it was close enough to the sea, and the large boulders, where they met and socialised. One in particular caught my eye – she stood proudly on the very edge, looking gallantly out to sea, her white chest puffed out, her beak as colourful as a children's toy. She was truly magnificent. The sea was still full of floating birds, and I couldn't quite believe my luck. Victoria was thinking of ditching the rhododendrons and studying puffins full time.

Joe took the time to show us where a white-tailed eagle had been nesting at the top of a steep gully. He used binoculars to try to spot the bird, but without any luck. I mean, it wouldn't have been difficult to spot with the sheer size of them. The nest had previously belonged to a golden eagle pair but had been used for the past couple of years by white-tailed eagles, although Joe didn't think that a chick had ever been successfully reared there, which was a real shame.

We made our way over to the isthmus between the islands. A small crowd of five or so people had gathered on the beach of grey pebbles, but Joe first secured the RIB to a well-used buoy, then deployed a tiny red plastic rowing boat from the rear. I gingerly eased myself from one boat to the other, feeling the instability of the rowing boat beneath me and fearing that a dip in the Minch would mean soaking wet clothes for the rest of my trip, and, worse, my notebook and notes being ruined forever. I managed it though, as did Victoria, and Joe got us safely over, where we were helped ashore. I was hoping that Adam Nicolson would be there to welcome me and show me around, but he wasn't, and I was ever so slightly disappointed. Joe was busy meeting and greeting people on the shore, and another smaller boat of researchers had just headed out towards the sea arch.

The spit between House Island and Rough Island was a spectacular spot, only just over a metre wide at its narrowest point. The sea lapped at both sides. I stood on crunchy round pebbles that were peppered with black, rotting kelp. Rough Island reared up sharply from the shore, its fierce face towards me, but I could see its gentler slopes behind. I noticed a path leading up off the beach and decided to give it a go. It led to House Island, location of the only shelter in the Shiants. The cottage sat below a small cliff, facing out to its own delightful little cove, which was shallow and had a shingle beach. Four long-tailed and hardy-looking sheep roamed the beach, like a family on holiday in Blackpool. The single-storey cottage had a chimney at either end, a doorway in the middle and windows either side, as if a child had drawn it. It had a smart, red, new-looking, corrugated-iron roof, and a faded and peeling blue door with a single piece of weathered wood on it bearing the legend 'Shiant Isles'. Two wooden chairs sat outside, made out of dilapidated rowing oars, and the mug on the window ledge was held together with duct tape. The makeshift table outside clearly served as the washing-up area – there was no running water at the cottage. Washing-up was catered for, drinking water came from a nearby shallow well, and the toilet? It was suggested that visitors should go to the toilet in the inter-tidal zone. Good luck with that in the middle of the night.

I pressed my face against the glass of the kitchen window and peered inside. It looked warm and homely; the fire was lit and there were wooden chairs dotted around. The cottage is used as a base for the research teams that come out to the Shiants – for socialising, cooking and warmth – although they often sleep in the tents that I saw dotted around the cottage in the 'garden' area. Some buoys had been fashioned into a rudimentary puffin. Judging by this, there wasn't a lot to do around here when the weather closed in; being stranded at the cottage while winter storms ravaged the islands would be pretty terrifying, I'd imagine.

I explored House Island for a while, never straying too far for fear of being left behind. I wouldn't survive very long – I'm hardly Bear Grylls. It was a wonderful, intriguing little place, full of nooks and crannies. Each time I turned a corner, I was greeted with another fantastic view of mountains, islands, skerries and seascapes, and not another person in sight. There were an estimated quarter of a million seabirds on the Shiants, and the sheer number of them was exhilarating. I spotted some little terns perched on rocks in a secluded bay, and great skuas patrolled the islands like a hired but slightly dodgy security firm.

Around the islands, I noticed signs of the Shiant Isles Recovery Project. Similar to the project I'd seen in Lundy, this was a joint undertaking between RSPB Scotland, Scottish Natural Heritage and Adam Nicolson's family to eradicate the black rat from the islands. Every now and then I spotted an incongruous yellow plastic container in the undergrowth – baiting stations for rats.

The fightback had started. The Shiant Isles Recovery Project proudly declared the Shiants rat-free in 2018 and, since then, storm petrels have returned, Manx shearwater numbers have increased and it's hoped the puffins will benefit too.

I was lucky enough to meet a couple of members of the Gortex-clad RSPB team, who told me that out of the fifty-five puffin burrows they were monitoring, forty-four had a puffin egg within – significantly higher than previous years. Even the oldest recorded puffin – at the grand old age of thirty-four – was found on these remarkable islands. The more typical lifespan is twenty years. Thousands of puffins depend on the Shiants to breed; let's hope they stay. A similar project, albeit with a significantly smaller population, south of the Shiants, at Canna in the Small Isles, had resulted in the puffin population doubling to nearly two thousand for the first time in a decade, and I'd seen the impact on Lundy and Anglesey too.

I made my way back down to the shore, as Joe was shouting for me. He had already shipped the camera crew back out to the RIB, along with their mounds of equipment. The RIB was sitting considerably lower in the water as a result. Victoria and I were the last to be collected in the small red boat. She went first, and then I followed. Well, I tried to. You had to time your steps down the beach with the outgoing waves, so that you could get close to the boat before stepping up and in. I completely mistimed this, meaning I stepped up and into the freezing cold Minch. I gasped out loud as the water filled my boot. I'm sure Victoria smirked to herself. I put on a brave face in front of my new crewmates as I scrambled back aboard the RIB. There were now six of us, including the RSPB volunteers I'd met earlier, and Joe was keen to get us moving as quickly as possible. He said that there was a storm coming in, and besides, I was worried about getting hypothermia in my left foot. Joe told us to hang on tight and floored it all the way back to Stornoway. I kept my eyes tightly shut as the boat lurched from swell to swell, hitting the water as if it was concrete, over and over again.

17

Pilgrimage

July

Bardsey/Ynys Enlli, Gwynedd, Wales

I was making the pilgrimage to Bardsey to see puffins, although I was concerned that this might be a last-ditch effort. It was late, late July and puffins would soon be leaving Bardsey, and other puffin enclaves, for the winter. They tend to leave in one go, suddenly. It feels like one day they'll be there, the next they've gone. They'll then stay exclusively on the waves for the following eight months or so, fishing for food when they need it and depending on their waterproof seal of feathers to keep them warm and dry. They even tuck their feet beneath their feathers to keep them warm from the freezing seas. Pufflings, on leaving the relative safety of the burrow for the open sea, won't come back to land for two years or more. They grow up at sea.

More amazing even than that is the range of destinations to which they go. Unlike other birds, which all migrate from or to a single location, puffins don't follow such rules. Puffins from UK colonies fitted with GPS geo-locators have been tracked as far north as the coasts of Greenland and Iceland, and as far south as the France and Spain. They seem to make their own pilgrimage to where food is most plentiful, but it's not really understood how they know this, or why individual puffins make the same trip each year.

These tiny birds brave the fierce winter oceans alone, bobbing around on the sea and seemingly thriving in the most difficult of conditions. That's pretty remarkable. It's not really known how they find their way back to their breeding ground. Theories include that they use scent, navigate by landmarks or are genetically pre-programmed. Puffins are true birds of the sea, salty master mariners only forced to come to land to breed. I just hoped none of that had kicked in for the puffins of Bardsey just yet.

It was a hot, muggy Saturday and the roads to Aberdaron in North Wales were clogged with caravans and tractors. In Kent, traffic was queuing for fourteen hours to leave the country. Kids were playing football on the M20, and the coastguard was delivering bottled water to dehydrated drivers. It was not quite as bad in the leafy lanes of the Llŷn Peninsula, but it was busy. Aberdaron was snarled up, and the families buying rock-pool nets and ice creams were fighting for space with cars driven by dads who were sweaty and worn out and fed up.

Bardsey is about two miles off the end of the Llŷn Peninsula and just sits there quietly. It's unassuming. I'd been visiting North Wales for years and had never heard of it. It might be only two miles away, but the seas there are a dangerous swirl of violent currents, giving the island its Welsh name of Ynys Enlli – Island in the Ferocious Current. Colin, the Welsh-speaking boatman who pilots boats between Porth Meudwy near Aberdaron and Bardsey, frequently has to cancel trips across as a result. This was my fifth attempt, including one which had ended with a miserable takeaway pizza in a thunderstorm in nearby Porthmadog. Even the pizza tasted soggy.

I was glad, therefore, to see Colin waving cheerfully from his little yellow boat as he rounded the corner into the tiny, sheltered, natural harbour at Porth Meudwy. There was a simple concrete slope that tipped into the sea off a shingle beach. It is National Trust property, and the car park was some distance away, up the emerald-green hill. There was no ice-cream shop here, and the crowds of Aberdaron seemed much further than only a mile away.

Colin was wearing a checked shirt, chest-high green waders and a ready smile. He proceeded with his slightly unusual way of getting passengers on and off his boat. On arriving in the harbour, he skilfully manoeuvred his boat between two metal poles sticking out of the sea. He eased the boat slowly towards the shore, as he'd done a million times and, once in position, hopped off it and into the seat of a waiting, rusty tractor. The tractor started with a puff of diesel smoke, and Colin backed it up the slope. The boat rose from the sea, neatly sitting on a trailer with wheels (and metal poles for sides). Colin flipped down a ladder on the back, and the boat was ready. He then unloaded both passengers and cargo from Bardsey, before shouting a spirited 'all aboard'. About ten of us scrambled up the ladder and on. There was a wooden box in the centre of the deck and a couple of metal lockers either side. We used these as seats.

'Keep all elbows and protuberances in the boat,' yelled Colin as he jumped back on the tractor and lowered the boat once more into the calm sea. He was smiling from ear to ear as he rejoined the boat and spoke to each of the passengers in turn. 'Edward,' he said, 'so glad you made it this time. I was beginning to think you were mythical.' I smiled and chose not to mention wasted trips and soggy pizza. I was genuinely pleased to be aboard and couldn't wait to get ashore on Bardsey.

We edged out into the sea past dark, mysterious caves that were just short of a pirate or two. Once we were out of the bay, the sea turned choppy, and plumes of spray jetted up the sides of the boat, as Colin guided us around the tip of the peninsula and towards Bardsey.

Bardsey looked like a typical Welsh mountain that had been dropped accidentally into the sea. The mountain – Mynydd Enlli – dominated the north of the island and its eastern side was almost sheer and sparsely covered with grass. It was on these steep and treacherous eastern slopes that the puffin colony had its home. As Colin told us this, I realised that I'd forgotten my climbing ropes and crampons, and my heart sank. Not only was I visiting Bardsey on a stifling hot day, in the afternoon, in late July, but I might not be able to see any remaining puffins anyway. Not without risking my life.

Colin steered us into the only landing spot on Bardsey, at Cafn, a narrow, natural harbour, and repeated his complex manoeuvres with another waiting tractor, before lowering the ladder again. He handed us a map of the island and advised us to shut all gates and to get to Bardsey's only café soon, as it closed at 1pm. It had already gone midday.

I headed up the slope and passed the rustic boatshed, with boats resting on the grass. From there, I joined a rough, pitted, single-track road with high, thick hedges that hid fields of tall grass already going to seed. Again, I was struck by how this felt like a chunk of Wales. The vegetation was green and well established. It clearly rained quite a lot here. The farmhouse Ty Pellaf was still selling coffee and teas, and, oddly, bags of salad, but I ignored the handwritten slate signs. I had a packed lunch and, besides, I only had four hours on the island, and elusive puffins to find.

My plan was to head straight through the village to Bae'r Nant, a bay at the north end of the island, before turning east and up Mynydd Enlli. I stopped briefly at the schoolhouse, which has been used as a community centre since it stopped operating as a school in 1945. The door was open; I guessed that most on the island were. Inside, it was pretty much still a schoolroom, but now with wall-mounted exhibits about the island, a small library, and a phone with an old-fashioned rotary dial attached to the wall. The wooden floor was still marked with the scratches of frequently moved chairs. Above the fireplace, which had, without doubt, kept scores of Welsh children warm through cold, wet winters, was a stone that was uncannily like the head of a fox, and there was also a cloak hanging up which a local woman had made from wool collected across the island. Out of the windows, the lighthouse on the south could be seen perched on the flattest part of the island. I wondered how many kids had daydreamed through the same windows before feeling a blackboard rubber hit the back of their heads.

Next door was the Bardsey Bird Observatory at Cristin. I was hoping to gain some more information on the local puffin colony from the observatory staff, but there was no one around, just an honesty shop selling everything from Bardsey T-shirts to cans of Bass shandy. Sitting in the sun and drinking this weak alcohol seemed like an option, if just to relive my early teenage years.

The village consisted of a few more houses, another honesty shop and exhibition, and one of those awful compost toilets that always remind me of music festivals and leave me appreciating a sink, soap and hot water. I'm obviously getting old.

Further towards the bay, I came across the cemetery and abbey. Folklore would have it that twenty thousand saints are buried upon Bardsey and, as unlikely as this may be, Bardsey was popular with pilgrims in the Middle Ages and remains so even to this day. It was commonly believed that three pilgrimages to Bardsey, with its rough seas and tough crossings, equalled one pilgrimage to Rome. The abbey was now in ruins, however, which seemed a shame seeing as it was once held in such high regard; just a single Celtic cross stood proud. In Iceland, puffins are sometimes named *profastur*, which means preacher. The puffin's Latin name is *Fratercula arctica*, with *Fratercula* derived from 'friar' or 'little brother' after the similarity of puffins to friars wearing robes. It struck me that it was entirely appropriate that I was making this pilgrimage to the isle of twenty thousand saints to find the little brothers.

Nearby was a cute little white house with a red door and red window frames. It was tiny, the sort of place that would feature in a fairy tale. It had a few shrubs around it and a grassy path leading up to the door, currently occupied by a family of white goats standing there indignantly. I was interested in this place because I'd heard it was a favourite of one of the island's more unusual residents. The little owl had adopted Bardsey as its home, and there were now several pairs on the island. Despite my best efforts, I couldn't spot any. But I liked the thought that, from some little hole in the ground somewhere, a pair of bright yellow eyes in a thoughtful, frowning face was staring at me. I was just sorry I didn't get to lock eyes with such a beautiful, inquisitive little creature.

The bay was also quite beautiful, and I noticed that a few of my fellow passengers had ended up there and had already broken into their packed lunches. This had not escaped the local gulls, which were slowly flocking around them to grab a crust of Hovis.

It was not far from here that the Bardsey apple was discovered, as recently as 1998. A birder staying at Cristin was enjoying apples from the old apple tree next to Plas Bach farmhouse. The apples were small but tasted surprisingly good. The birder thought so much of them that he sent a few to a fruit-growing friend, who had never seen the variety before. The apples were later declared their own distinct species, not found anywhere else. The Bardsey apple has since been reproduced around the world, but the original tree remains on Bardsey, producing its delicious, unique apples. I love the whole story of the Bardsey apple and hope to one day crunch into one.

I headed eastwards, through a patch of bracken and a small copse of conifers. The path here was indistinct, covered by gorse and fern, but I pushed on through, watched by various sheep. As I gained height, some of the vegetation thinned away, leaving the bones of the mountain on show through its thin skin and flesh.

The sea stretched out in front of me, calm and clear in all directions. A few stunted trees reminded me that conditions there weren't always quite so benign, and that it was surely a fearsome place in the winter months. The coast is jagged and gnarled, thrashed by the wild Welsh seas and wind. Wheatears and stonechats flitted by, accompanying me up the steepening slope. I even caught sight of a rock pipit and, on the ground, I noted the plentiful burrows of Manx shearwaters.

Near the summit was a flat area of grass, surrounded by heather and thistle. It had become the play area for four or five choughs, who were jumping around excitedly, sounding their sharp caws and displaying their cartoonish red beaks that give them away so easily. The collective noun for choughs is a chattering, which is entirely fitting. King Arthur is said to have left this world as a chough, which seems appropriate with his links to Bardsey; some folk tales will have you believe that the King was buried on the island. I stopped and watched the choughs for a while, starting on my warm cheese sandwiches from their foil packet and trying to give them some texture and taste by inserting crisps between the soggy slices of bread. At least the choughs gave me plenty of entertainment.

From here, the path became more obvious and began to run parallel to the east coast. This felt very mountainous, with sheer slopes that rolled directly into the sea below. They were mainly grass covered, but I could see burrows across them. I hoped the ones near the bottom were puffin burrows, but I could see only herring gulls patrolling the lower slopes, and cormorants right on the shoreline itself. I hoped I hadn't missed them – there was an unmistakable silence where there would recently have been the chatter and laughter of masses of auks and, in places, the white marks of guano finished suddenly, as if painted against masking tape, where shelves of guillemots and razorbills had sat. I strained my eyes to see if I could pick out the distinctive shape of a puffin amongst the few remaining birds. I couldn't.

The path climbed higher, taking me away from the birds, before eventually dropping again and offering views over southern Bardsey, with its red-and-white lighthouse standing guard. Tiny fields covered the island, separated by thick green hedgerows like a crocheted blanket made from only green wool. I tried to find a path that would take me eastwards, back around to where I thought the puffins would be. I could see a brown dirt path running a few metres from the sea, but I wasn't sure if it was a sheep track or whether it was safe. I didn't want to end up slipping into the sea. Further down, a wooden sign perched against a stone wall asked that no one pass that way between March and September, to protect breeding seabirds. I retreated back down the hillside, seriously worried I wouldn't see the Bardsey puffins at all.

Back on the flatter, less windy southern half of the island, Pen Diban, I could hear an eerie sound coming from the bay at Yr Honllwyn. It drifted across the island like a siren's call, albeit punctuated with the snarling and growling of a rabid dog. In fact, I initially thought it was a dog stuck in a fence but, on further investigation, I could see it was coming from the shallow bay.

The grass gave way to a series of boulders and shingle, met by the calm water and a band of rotting, brown kelp. Further out were fifty to sixty grey seals. Some were posturing on rocks poking out of the water, heads and tails pointing skywards. Others thrashed and riled themselves in the water, heads breaking the surface with slurps and gurgles. Grey seals have beautiful wet, wide eyes on their dog-like faces, giving them an inquisitive, surprised look, which was apt, as they were both inquisitive about and surprised by my presence. It was the noise, though, that really grabbed my attention. It was a sad wail, but accompanied by a sound that could be used on a horror movie. Maybe it already has been, in *The Predator*. This was the build up to the breeding season, with the males noisily protecting potential mates while, at the same time, ensuring a captive audience. It truly was like nothing I'd ever heard, a whining, yearning cry that was quickly caught on the wind.

It was time to board the boat. Colin was only slightly late, but I was grateful to see him at all, especially when he mentioned that a change in the weather meant there were unlikely to be any trips tomorrow. I liked Bardsey, but I didn't fancy being stuck on the island, especially with the seas howling through the night and having to wait for the café to open. I suppose I could have survived on Twixes and Walkers Crisps from the honesty shop, and maybe the bird observatory would have let me in so I could sleep in front of the fire like a dog. Either way, I much preferred to see Colin repeat his peculiar way of heaving his boat ashore, and I gratefully clambered up the short metal ladder. I was particularly pleased as Colin had also promised me puffins. If there was one man who could make good on such a promise, it was Colin.

We set off, keeping all protuberances safely inside, and launched into the sea in the style of the RNLI. I took a smack of salty, cold seawater to the face. Colin headed north-east, around Pen Cristin and towards Ogof Barcut, which I'd been directly above earlier. I could just make out the path that I'd walked and the ledges now empty of guillemots and razorbills. There were still a few of both varieties on the water and on the cliff faces, but Colin reckoned 70 per cent of the birds had left in the space of a week.

We continued moving north up to Ogof Braichyfwyall. I scanned the steep slopes for signs of puffins. I could see burrows and the bare dirt where their spring exploits had scratched away the delicate foliage. There was a shout from the boat. Two young girls had spotted puffins to starboard. I craned my neck and briefly thought about pushing the two under-tens aside to get a better view. I didn't need to. Colin slowly turned the boat, allowing us all to get a view from the stern. In front of us were around twenty of Bardsey's finest puffins. They were truly beautiful, and so calm. They were floating on the gentle, milky-green sea, unperturbed by our presence. They had bright, clean beaks and sugar-white cheeks. Every now and then, as the boat floated close to a puffin, the bird would fly up and along the surface before flopping gently back into the sea a few metres away. There were no puffins on the shore, I noticed, but that didn't matter. I was so pleased to see them at all.

One bird in particular was very curious, constantly looking at the yellow boat with several human beings pointing and staring back. I suspected he was a younger bird. Puffins three to four years of age sometimes return to colonies at the end of the breeding season to look for mates, or consider burrows, or just to generally get a feel for the area. I guessed that this was exactly what this little chap was doing. I focused on him, watching him gently floating by. It would have made my day to have just seen him; I wouldn't have minded if he'd only stuck around for a few seconds, but he was putting on a show, dancing around the boat and showing his wonderful beak to anyone watching. I was thrilled. I'm not religious, but my own pilgrimage to Bardsey was now complete.

18

Last Orders at the Puff Inn

July

St Kilda, Outer Hebrides, Scotland

The Minch was silver and silent. The sea barely moved and could not be distinguished from the sky. It was early morning, and the sun had yet to make its mind up, instead illuminating the spectral scene from above. Harris appeared as a smudged written sentence across a grey page. The Fladda islands were just full stops.

St Kilda is not the easiest place to get to or from, a rugged archipelago some fifty miles west of the Outer Hebrides themselves. This was one of the reasons for its decline and subsequent evacuation in 1930, when its last thirty-seven residents had had enough. This isolated community had impossibly hard lives, compounded by illness and, in the last years, a shortage of food. The young and able had already departed for better lives in Australia and America. Fearing another winter on the island, the remaining inhabitants decided to leave once and for all, and so, on 29 August 1930, they boarded HMS *Harebell* and were evacuated to the Scottish mainland.

I boarded the *Integrity* at Uig Pier, off the Isle of Skye, a neat little boat with a covered section and comfortable seats that would see us make the four-hour trip to St Kilda come what may. I'd heard horror stories about this journey, mainly involving vomit, and had already had two visits to the islands cancelled due to the weather. It was now late July and I had weighed the risk of missing the puffins against the risk of being able to make the trip at all, or at least while retaining the contents of my stomach.

Razorbills and guillemots rested on the serene sea, only to dart away from the boat as we gently cut our way through, waves falling from the bow with ease. The birds left little wrinkles in the surface as they frantically ran across the water before take-off. A bottlenose dolphin joined us for some of the trip, riding the bow wave and eliciting exclamations from my fellow passengers. This was only part of the surprise though, as a baby dolphin joined her mother. Only a quarter of her mother's size, and still with the fold visible across her flank from her time in her mother's womb, she too was enjoying the wave, repeatedly leaping in the soft light and landing with a small splash. Her light underbelly was the same colour as the silver sky. It was a wonderful sight and a reminder of the magnificent, intelligent creatures with which we share the world.

We made one stop, on the island of Berneray, to pick up a couple more passengers. Berneray looked idyllic and I put it on my list of places to visit. I could see the multicoloured wildflowers of the machair beyond the wide, white beaches and, even at that early hour, the local children were taking it in turns to jump from the pier into the crystal-clear sea. Looking down, I could see pink sea urchins on the sea floor, like dropped bath toys.

We drifted past the lonely islands of Pabbay and Shillay. Later, we passed Haskeir, which was nothing more than a slab of rock that has somehow managed to support a lighthouse. Haskeir was known as Skildar on Old Norse maps and may be how St Kilda got its name, Skildar having been confused with the islands to which I was heading.

Once the boat was out on the open North Atlantic, most of the passengers and crew came in from the viewing deck. The sea was now noticeably rougher, and Andy, the skipper, was clearly trying to make up time from the Berneray stop. The result was that the boat skipped from swell to swell, causing my head to nod involuntarily and violently and making my internal organs feel as if they were becoming detached one by one. At times, it seemed like I was being slowly but determinedly beaten. I politely turned down a coffee.

Nicola joined me. She reminded me of a young Julia Bradbury from *Countryfile*, with her brown ponytail and big brown eyes. She had recently got married to Andy, the skipper, and was from the well-known seafaring town of Nottingham. She'd been doing this trip for eleven years though, and I took some comfort from the fact that Andy was unlikely to want to kill or injure his new bride. Nicola told me that it was actually pretty calm today, and then regaled me with tales of whole boats being plagued by non-stop vomiting and a story about the time when they were trapped on Hirta as the weather wouldn't let them home. You know, just to quieten my nerves.

The ecosystem on St Kilda is unique. Home to nearly one million seabirds during the breeding season, this Natural and Cultural World Heritage site is remote enough to carve its own ecological path. The St Kildan sheep, mice and wrens are all evidence of this; some are getting bigger in order to survive – wrens, mice – while the sheep are getting smaller. It's a standing joke that, at the current rate, the mice will soon be chasing the sheep around. This also means that the islands have to be closely protected. Which is why boats are not permitted to dock at Hirta. Instead, visitors must be transferred by dinghy to cut down on the chance of rats or mice catching a lift ashore. While I fully understood the need for this, this business of transferring between boats was my least favourite part of all my puffin trips, not helped by my having nearly lost a boot in the Shiants. But imagine coming all this way and refusing to board the tiny, damp dinghy that would take us ashore. So, as we approached Village Bay on Hirta, I reluctantly swung my leg over and perched precariously on one of the gunwales, wrapping my fingers under wet rope handles.

Hirta is the largest of St Kilda's islands. Since 1930, it's remained pretty much as it was, and Village Bay felt like a film set. I made my way down to the beach, after the obligatory safety briefing, and looked out. The sharp dragon's teeth of nearby Dùn rose out of the sea to my right. Sea cliffs were to my left and an old artillery gun still pointed menacingly at the bay. The bay soon gave way to lush, green slopes of grass, which eventually fell away into the sea from the tallest sea cliffs in the British Isles. I was struck by just how remote and isolated this place really was. Four boats, including *Integrity*, bobbed gently in the bay. The wind whipped at my face and caused an oystercatcher to hurry diagonally across the white sand, leaving tracks as if it were snow. I headed on into the village, managing to stand in a fresh pile of sheep shit on my way, something that you don't often see in the films.

There's been a military base on Hirta since 1957, with the Ministry of Defence renting its space from the National Trust for Scotland and bringing power and running water as a bonus. It must be a lonely old posting, and challenging too; in 2015, the weather was so shocking that all personnel were temporarily withdrawn. That's how bad the weather gets on St Kilda – it makes the army move out. The army has a sense of humour though – you'd need one, living on the remotest, wildest islands in the UK; its canteen on Hirta used to be known as the Puff Inn, although, sadly, it served its last drinks in 2019.

Near the military base I found the Manse (now with a small museum, a shop and the most pristine toilet facilities I'd ever seen on an island), the schoolhouse, the church and the remains of the rest of the village. There were also a couple of jarring modern temporary buildings just down from the village and an alarmingly blue tennis court that I suspected didn't see much play. The narrow main street was dotted with clumps of rampant stinging nettles and lined with a curved row of stone 'blackhouses' looking out to sea. Some of these had been lovingly restored with new roofs, while others had been left to decay. Whole families would live in these, in a single room, often with their cattle at one end and a roaring fire at the other. Number 2 in the street had a slate outside stating that it had been the McDonalds' house, passed from John McDonald Senior to John McDonald in 1883. Intriguingly, it was also known as Ladies Boudoir, so goodness knows what John got up to.

Also dotted around the village were small stone huts with turf roofs that looked like they needed a haircut. Known as cleits, these were used by St Kildans to store and dry everything they needed, including the harvest of seabirds and their eggs. Most of the cleits were still standing, and in many I found a surprised Soay sheep taking shelter. With the seas often too rough for fishing, seabirds were a staple meal for the St Kildans, with one census in 1764 showing that each St Kildan would apparently eat up to eighteen seabirds and thirty-six seabird eggs a day – a difficult concept for us to grasp, with seabirds now perched so precariously on the edge of decline. Gannet, guillemot, fulmar and puffin were all popular, either fresh or salted and preserved in cleits. Fulmars were once only present on St Kilda and nowhere else in the UK. They were seen as a true commodity, with their stomach oil used and sold as a cure-all medicine, while their down was exported to the mainland for bedding.

Dogs were used to hunt puffins in St Kilda, just as they were in Norway. Known in Norway as Lundehund, which literally translates as puffin-dog, they were specially bred to hunt puffins, with extra claws for holding on to steep cliffs and ears that would prevent loose soil from entering when raiding underground puffin burrows. In St Kilda, puffins were also taken with large nets on poles, using techniques similar to those still employed in the Faroe Islands and Iceland, as well as more commonly being caught by placing multiple snares on rocks on to which the puffins would land, or single snares for taking individual unwitting puffins. Puffins were plentiful and notoriously easy to catch by the St Kildans.

Seabirds are still eaten in one place in the UK. A small group of men sails from Ness on the Isle of Lewis to the remote outpost of Sula Sgeir each summer to harvest guga, juvenile gannets. The men are permitted by law to take up to two thousand of the odd-shaped chicks each season. After scaling perilous cliffs, the men spend up to two weeks on the island, taking guga from their nests. Swiftly despatched with a wooden pole, the guga are plucked and any remaining feathers are removed over a roaring fire. They are then heavily salted and placed in huge circular stacks awaiting transport back to Ness. The guga are taken to be eaten by the hunters and their families, but any surplus is sold on to the queue of locals that form to meet the returning boat. Surprisingly popular in Ness, salted guga is cooked and served with boiled potatoes. In his excellent book, *The Guga Hunters*, Donald S. Murray jokes that guga may not be exactly tender to eat: 'Place a guga and a stone in a pan of water and boil. Once you can pierce the stone with a fork the guga is ready for eating.'

The hunt remains controversial though, to the point that guga hunters are under increasing pressure to stop the tradition – including threats being made to their lives.

St Kildans tucked into puffin meat at every opportunity. To liven up a breakfast porridge, they would add puffin meat to it, something that they leave off the Quaker Oats boxes these days. Some communities, including in Iceland and the Faroe Islands, still dine on puffin meat. In Iceland these days, the hunting of puffins is limited to a short window of days to manage the number of birds that are taken. Not all Icelanders relish the taste of puffin and, in recent years, it has fallen off the tourist menus in Reykjavík. I was once invited to the house of an Icelandic friend who had taken a great deal of care in procuring puffins and preparing a dish for me. Following a recipe handed down through his family, he had carefully plucked the birds and removed the dark breast meat. The carcasses were boiled down to a stock with onions and carrots, to be spooned over the sliced breast meat, which was served rare but warm. It tasted of game, with a slightly fishy tang and, although not unpleasant, I had to wash down each mouthful, and my feelings, with a gulp of red wine. I'm not sure I could repeat the feat.

I bumped into Nicola in the cemetery, where gravestones stood as testaments to the short, hard lives that had been lived on the island. A tetanus epidemic had killed two thirds of all new born babies on Hirta, which I found extremely sad. Nicola explained that the cemetery had been built without corners to prevent the devil from hiding, but I couldn't help thinking that it wasn't the devil the St

Kildans should have feared. A meadow pipit dropped by to cheer me up, and was quickly joined by a wren. The St Kilda wren is also different to its mainland cousins. It's bigger and has a longer beak – often pointed skywards – and more stripes than the tiny 'Jenny' wrens, as my gran used to call the ones in her well-tended English garden. I was grateful for the distraction of this little bird. It danced from rock to rock, ever alert, bobbing up and down with its jaunty upright tail and surprisingly loud call, unaware of the sadness of the cemetery.

The sheep, mice and wrens weren't the only animals to have adapted on St Kilda. Nicola had had the pleasure of meeting some descendants of native St Kildans, including a chap whose great-grandmother had lived at Hirta. Nicola got him to remove his socks and shoes to reveal a pair of small, wide feet and thick ankles perfect for scaling cliffs to collect birds' eggs. He had always struggled to understand why he couldn't find shoes to fit.

Hirta's best-kept secret is its perfect beach of white, smooth sand leading into the pristine, turquoise sea. The sands all but disappear in the winter, only to return with the tide each spring. I sat amidst the iris beds as a visiting whooper swan honked somewhere nearby. Other recent visitors had included a snowy owl, and for the first time, swallows had started nesting in the military buildings, causing the army to keep their doors wide open for the summer. Fulmars patrolled the skies like an aerial taskforce, while a handsome ringed plover skipped across the sand.

I walked out of the summer pastures and up onto the slopes, first as if heading towards the cliffs of Oiseval to the east of Hirta, and then onwards, to Conachair. Looking back down on the village, I noticed that the buildings and cleits, walls and pastures were neatly outlined in near concentric semi-circles, but surrounded by clusters of stone that had fallen away as if built messily from children's building blocks. The crofts had once been fertilised with the remains of seabirds, as nothing was wasted by the St Kildans. I wondered how many thousands of unfortunate puffins had ended up this way. Conachair is the highest point of Hirta and was worth the walk, despite the great skuas, who were being their usual unwelcoming selves. From there, I could see the grassed slopes of Soay, original home of the Soay sheep, and, in the distance, the fierce islands and sea stacks of Boreray, Stac Lee and Stac an Armin, their tops white with gannets. Stac Lee and Stan an Armin stood either end of Boreray as if for protection from the feisty Atlantic. Near to where I stood were the remains of an aircraft that had been downed in the Second World War and was now slowly sinking into the sphagnum moss, just like my feet. There are two other two wartime plane crashes, including one on Soay.

Boreray was a quick four-mile hop across the choppy ocean in the *Integrity*. Boreary has steep, grassy slopes on one side and towering and angry spires of rock on the other. Looking closely, I could see the remains of cleits, and even the small bothies that were used right up until the evacuation as temporary accommodation for those brave enough to tend their sheep or gather gannets from the imposing Clesgor cliffs. Since humans had stopped visiting Boreray, great skuas had returned in abundance.

Stac Lee looked like a giant fir cone, with guillemots and gannets marking each diagonal ridge. I struggled to focus fully on the island, as the swell was causing the *Integrity* to rise and fall. Guillemots surrounded the boat, seemingly not experiencing the same problems as me.

Stac an Armin, the tallest sea stack in the UK, loomed out of the sea like an evil lair in a fairy tale. It towered over us, one side illuminated by the sun, the other in forbidding darkness. Rather than dragons, though, it was gannets that were riding the skies here, emerging from the stack's own topping of cloud before plunging into the sea beside us. There is a story that a diving gannet once hit a ship in this spot, with such force that its beak penetrated a wooden plank. The noise of the gannets was overwhelming, echoing off the cliff faces, and the sea was both frothing with diving gannets and being bombed with guano. I kept my hat on.

Stac an Armin is the setting for a rather sad tale, as told in Errol Fuller's *The Great Auk: The Extinction of the Original Penguin*. In 1840, a group of men found a great auk on the stack and effectively kidnapped it, hog-tying its legs and taking it with them to a nearby bothy. An unrelenting storm blew in, trapping them all for several days. St Kildans were a suspicious lot, and these men were especially so, it seems. They concluded that the great auk was a witch – it was obvious, wasn't it? – and had caused the storm. There was only one solution. They beat the last surviving great auk in the UK to death with stones. It's not clear whether this caused the storm to stop.

The great auk was flightless and struggled to walk well on land, instead spending its time awkwardly waddling around. Much like its surviving razorbill and guillemot cousins, the great auk was much more at home underwater, where it could dive to great depths in pursuit of fish for food.

Over the centuries, every part of the great auk was sought after; meat and eggs for food, feathers for bedding, its rich oil for lighting and heating, and its tough skin for clothing. As if that wasn't enough, the Victorians were prepared to pay high prices for a great auk, with the upper classes wanting nothing more than a stuffed specimen to sit on their writing desks. In addition, it was ridiculously easy to catch. There are reports of sailors placing planks of wood between their ships and the land, then simply rounding up whole colonies of the big daft birds, who would merrily walk the plank, in reverse, to their demise. No one is saying that they were particularly intelligent.

The last ever great auks were said to reside on Eldey, a remote island off the south coast of Iceland. In 1844, they had the misfortune to be found by three Icelanders, who later admitted to strangling the auks to sell in exchange for hard cash. At nearby Reykjanestá, a copper statue of this lost bird looks out to Eldey, standing in quiet memorial. I have been to pay my respects.

Isn't that the saddest demise? This once great bird was taken forever from the world because of the actions of greedy, destructive men. All I can hope is that we learn something from this sorry episode, and that the puffin – or any other species, for that matter – doesn't go the same way as the poor great auk.

Pleasingly and reassuringly, Dùn, the third St Kildan island, proved to be absolutely covered in puffins. St Kilda is the UK's largest colony of puffins, with nearly 142,000 pairs, all of them now safe from being harvested. Most of them seemed to be on Dùn. From the boat, Dùn seemed to be wrapped around Village Bay like a serpent's tail. Guillemots lined the narrow ledges next to the collapsed sea arch of Dùn Gap like ornaments on a mantelpiece, but it was the puffin high-rises of Dùn's steep slopes that drew my attention. Here, puffins stood guard outside their burrows in their thousands, while the deep blue sea below was full of countless other puffins diving for fresh fish and returning to the surface with shiny mouthfuls of sand eels. Each time the boat bobbed in the gentle swell, another raft of puffins appeared, all bright beaks and white cheeks, as if freshly washed.

It was a magnificent sight, and it made me think of Skellig Michael in Ireland and the filming of *Star Wars: The Last Jedi*. The story goes that the filmmakers arrived at their Skellig Michael location to find it covered in puffins. Unable to remove the puffins physically or digitally, the crew decided to keep them in, and later upgraded them to porgs. So, the pint-sized porgs in that film are actually digitally enhanced puffins.

The same thing could easily have been done at Dùn. Above us, the sky was black with puffins, like midges buzzing my head on a hot summer afternoon. They twisted and turned, returning to either the sea or their burrows, forever busy, feverishly fishing and feeding their chicks before the summer ran out. Puffins for as far as the eye could see.

19

Orkney Rock Stars

July

Westray, Orkney, Scotland

The rain at Kirkwall airport was unbelievable. It was late July, but it could have been late November. Stepping from the plane to the tarmac was like taking a shower, and I was instantly drenched. The poor schoolboy who had been stationed at the edge of the runway to welcome my Loganair flight – Scotland's only airline, don't you know – with a bagpipe serenade was dripping wet. Like his ancestors before him, though, he played on. His mum watched through misted glass, proud as Punch and with a towel in hand.

I caught the bus into Kirkwall itself, some four miles away. The rain lashed at the bus as passengers slowly gave off their own clouds of steam. At the bus stop, lines of elderly cruise-ship passengers queued for express trips to archaeological hotspots Skara Brae and the Ring of Brodgar, while a lady in a Visit Orkney caravan gave out leaflets in every language, simultaneously explaining that it would not be possible to get to Edinburgh and back before the ship left. In any case, the Tattoo wasn't on. I found my accommodation after asking around, handily placed next to a funeral director. To be honest, it suited my mood. The weather was bad enough but, on first appearances, Orkney hadn't endeared itself to me. It was flat as a pancake and Kirkwall was an uninspiring collection of slate-grey buildings with a rundown pedestrian street through its centre. I ordered a bacon bap, but it contained sinewy strips of uncooked pig fat rather than crisp, tasty meat, so I threw it in the bin. A pint was in order.

Down at the harbour, the boats and ships were being dragged around by a fierce wind, and the raindrops were being fired like bullets. Helgi's had a suitably Viking name and occupied a cute building between a couple of elderly harbourfront hotels. From the bar, I could see the RNLI station and the huge orange rescue vessel tethered nearby. Heaven help anyone who needed their services today. I ordered a bowl of cheesy chips – comfort food; the more calories the better – and, eliciting the first smile of the day, a pint of Orkney Brewery's Puffin Ale. It was good stuff too, dark and tasty, and didn't actually contain any puffins.

I eased myself onto a stool and began chatting to the not unattractive barmaid. Handing over a damp £10 note with an apology, I explained my strange mission. She smiled, clearly amused. 'You could try the Brough of Birsay,' she said, 'but there aren't many puffins on the mainland. You have to be quite lucky these days.' By 'mainland', she meant the Orkney mainland, not, you know, Scotland. 'Otherwise, it's Westray, where you can sit right next to them.'

She was called over by another customer, and I was left pondering over my beer. It seemed that a trip to Westray might be the best option. I asked her to point out the ferry to me, and after another pint and a fascinating fact – she was from Stronsay, which was exactly the same size as Ibiza but had none of the nightlife – I wandered over to get the information for tomorrow's ferry. The nearby butcher had a sign that read, 'Who gives a fajita about the weather?' and I chuckled to myself as the rain ran down the back of my neck.

Later in the afternoon, I took a local bus out to Birsay. I couldn't see out of the windows, and it smelt like hay and farmers' old boots. The rain continued. The bus didn't go through the village of Twatt, which was disappointing, and the driver chucked me out a couple of miles short of Birsay. I schlepped the rest of the way to find that the Brough of Birsay was a tidal island and the tide was coming in fast. I gave up and instead headed to a local garage to wait for a bus home. It was just past Earl's Palace, a sixteenth-century ruin of a palace built by the First Earl of Orkney. The lady inside gave me a cup of molten hot chocolate to fend off my pending hypothermia and said that she had seen very few puffins all season.

Seabird colonies at Marwick Head and the Brough of Birsay had seen massive reductions in the number of birds. According to the RSPB, the kittiwake population had declined by 90 per cent since 2000. There were 570 pairs in

2015; the same site at Marwick Head held 5,573 fifteen years earlier. In 2013, the RSPB tracked 360 nesting kittiwake pairs and yet only a single pair successfully reared a chick. Sickening, isn't it? Surveys in 2018 showed a continued decline of most species, including fulmar, guillemot and razorbill.

From what the garage owner told me, and the sad tales I'd heard about Birsay, it seemed like Westray was the place to go. I took the bus home to the funeral director and hoped that tomorrow would be much, much better. And drier.

The next morning, I found the sky without a single cloud. I hurried down Victoria Street, past the imposing St Magnus Cathedral, which looked much better in blazing sunlight, to the ferry port. I was catching the 0713 to Westray, and I couldn't wait. The waiting room displayed gloomy posters about the dangers of being a seafarer, which crustaceans you could and couldn't take for eating, and exhortations to 'Ban the bruck (rubbish)'. There were pictures of crabs, whelks and lobsters. It wasn't long, thankfully, before I was ushered down the green metal slope of the roll-on, roll-off ferry.

The ferry was much more modern than I was expecting, and resplendent in white paint. Inside, though, nothing appeared to have changed since the 1960s. There was a firm emphasis on the colour brown; everything in the passenger lounge, from floor and seating to ceiling, was a shade of brown. Once we were on our way, and with miserable Kirkwall left behind, I headed to the ship's lower deck to find the canteen. This turned out to be a rather grand term for what was essentially a serving hatch and some

Formica – brown, of course – tables. I ordered a sausage sandwich and sat down to eat it as the ferry chugged and burred all around me. I began to grumble to myself about the sorts of things people put in cheap sausages, this one being seemingly full of crunchy gristle, when I realised that I was actually chomping on part of my front tooth that had dropped out. I decided not to complain.

Out on the open section of the upper deck it was a really stunning day, with the sea beautifully calm. Not calm enough to soothe away my loss of a tooth, obviously, nor to prevent kids from now taking cover every time I smiled. The islands slid past – Shapinsay, Stronsay, Egilsay and Eday. Even the names made me grin. Some had wonderful white beaches, others had patchworks of crofts that looked like a green tartan. Shapinsay is a hangout for the rare hen harrier as well as being famous for Balfour Castle. I vowed to come back to Orkney and take my time island-hopping, maybe with a bicycle and a fishing rod, catching my supper from the sea as I went. Orkney has some of the best fish in the UK. Or maybe I could come back in the cold winter months, grab a cosy cottage and wait for the Northern Lights to show. This was soft and gentle sailing though, and the sun was warming my face. The ferry was nowhere near full, and there was plenty of room to move around. I was quite content here, watching the islands pass by.

Westray is kind of cross-shaped, with Rapness in the south and Pierowall roughly in the centre, looking across at its smaller neighbour, Papa Westray. The ferries sail to Westray three times a day from Kirkwall, taking about ninety minutes and often calling at Papa Westray too.

There's also a flight between Westray and Papa Westray; at just two minutes long, it's the shortest scheduled flight in the world, so they do the briefing announcement before take-off, and you get a certificate.

I really enjoyed the ferry trip from Kirkwall, but I could see it being a real pain in some cases. Everything takes extra effort. Westray's main village is Pierowall, but it has few services. The fish and chip shop only opens on Wednesdays and Fridays. To stock up on essentials, islanders go to Kirkwall, where there are three major supermarkets in a row. My ferry was filled with building supplies, including a wheelbarrow, and had been ninety minutes late leaving as a result. The ferries are extremely well run, but I'm not sure I could make a habit of it. I was fortunate to be making the trip on a glorious day, but on a stormy day in November because you've forgotten to pick up enough toothpaste? I'm not so sure.

The ferry backed into Rapness. This wasn't, by any means, a bustling port. There was a single building housing toilets and a waiting room, with a phone to call for the island's bus to Pierowall, and a 'Wanted' poster asking for the RSPB to be contacted if a corncrake was seen or heard. A sign pointed to a nearby house which seemed to double up as a tea room.

Westray is known as the Queen of the Isles and it was not hard to see why. It was stunningly beautiful as I started off on foot. My map indicated that Castle o'Burrian was a couple of miles away, and I had plenty of time. The sun was blazing down and the grass was a deep, lush green punctuated by yellow splashes of ragwort. There was part of me that wanted to put a bit of hay in the corner of my mouth, hop over the hedge and have a quick snooze. It was that sort of day. Highland cattle watched me pass with interest, peeking out from behind giant purple thistles. The field had been cut for hay, and it smelt warm and comforting, while lichen on the stone walls told of damper days. In places, fragrant dog rose was using the wall as a climbing frame. I turned a corner to see a single straight road in front of me that seemed to stretch the length of the island; there were a couple of farmhouses dotted either side, but not a single vehicle or person in sight. I carried on walking, lost in my thoughts.

After a while, I heard the noise of a rattling old vehicle approaching. It was a stark sound compared to the birdlife and the low mooing of the Highland cows. A battered Land Rover came into view, seemingly held together with binder twine and rust. It slowed and stopped next to me. The driver was clearly a local farmer, dressed in the uniform of farmers everywhere – blue overalls. He had a kind, weathered face. He said something to me, but I didn't catch it. Not a word. Must be the wind I thought, and leant forward. He said it again. It was embarrassing now. I didn't catch a word. Without a sigh or any sign of exasperation at all, he leant across the passenger seat and pushed the door open. He was offering a lift.

I gladly accepted. The floral garden-chair cushion on the passenger seat took me by surprise. It smelt of cows and hay, but not in a bad way. I hopped in. I still couldn't understand his Orcadian burr, and he was probably struggling with my Shropshire/West Midlands accent.

Through the medium of mime, maps and shrugging, I pointed out where I was going. He smiled politely and continued driving for about four minutes down the straight road. A big hand-painted sign had been placed against a footpath sign. It had a puffin on it and an arrow. That was good enough for me, so I thanked the kind farmer and headed off down the overgrown footpath. I have often thought of him since, wondering if he likes to tell the story of a mad Englishman bothering the local wildlife and not being able to understand a word he said.

The footpath gave way to something squidgier underfoot and the sharp spines of reeds scratched at my legs. The sea was back in view, in a gentle cove called the Bay of Tafts, with a single wind turbine on the opposite shore. A couple of seals were basking on the rocks, while a sanderling picked its way along the beach. I arrived at an old mill house, built in the 1850s to grind oatmeal, with its wheel still hanging on stubbornly. Inside, the roof had gone in places but, judging by the floor, it was a firm favourite with the local bat colony. A family of wrens played on the rusting remains of the wheel, hopping from side to side. At one point, the smaller balls of fluff lined up on a pole like a scene from a 1970s kitchen tile. A blue dab of devil's-bit scabious caught my attention nearby. I was keeping an eye out for the uber-pretty Scottish primrose, a rare beauty that I knew grew around Westray. Both of my grandfathers had been keen gardeners, and I must have inherited some of their enthusiasm. Or maybe it's just an age thing.

I headed on. Behind a small sea arch, I found three nesting black guillemots and an over-large fulmar chick with a punk-rock fringe. I'd never seen black guillemots so close, and they were stunning in their satin-black plumage, white wing patches and bright, letterbox-red feet. The inside of their beaks was the same shocking red.

A couple of cyclists had set up camp further around the bay, right next to a patch of clover with giant purple flowers. They were intertwined in their tiny tent, looking out at the view, and sometimes into each other's eyes. Their bicycles lay nearby in the soft grass. They didn't seem to like me being around, so I carried on. I left an inquisitive twite on a barbed-wire fence to keep an eye on them for me.

The path opened out, showing angelica poking through, with heads as big as cauliflowers, and then Castle o'Burrian came into sight. It was a squat-looking sea stack, with a gentle, green slope opposite leading to a rocky shore and a narrow strip of sea. The slope gave way to steep granite cliffs in sharp geometric shapes like stacked books, with a fringe of tufty grass overhanging it. The guillemots and razorbills had departed for the season, leaving just puffins. And what a sight they were.

On the stack itself, they were spread across its flat top like a layer of black-and-white jam. They were everywhere up there, parading proudly back and forth. I could hear them moaning and groaning to each other. There were a few in the cracks and crevices too, but these are mainly empty and still streaked with the previous occupants' mess. I headed for the slope, which was covered in browning sea thrift. At the junction of the slope and the sudden start of the cliffs, I spotted a gaggle of young puffins. There were four or five of them, and they'd discovered a hole in the cliff where they could pop inside, take a right turn and then pop out via another hole. They were loving this game, and readily hopped back up to do it all over again. They were also practising gathering grass in their beaks. I couldn't help but take a seat and watch this little performance. I spotted another puffin picking its way through red campion and yellow marsh marigold. It was a beautiful image. If it hadn't been for the midges, which were increasing in number and bothering me, it would have been perfect. Bloody midges.

I followed the footpath south. It was a little overgrown but otherwise well looked after. It was a roundabout route back to Rapness and by no means the quickest. It took me away from the sea and, without the sea breeze, it felt quite warm. The sun was beating down on my arms. It was on this path that something rather special happened. I heard my first, and only, corncrake.

Corncrakes are classed as endangered in the UK. In 1993, the Orkney Corncrake Initiative was set up by the RSPB. It was thought that there were fewer than ten corncrakes in Orkney at that time, and the initiative sought to reward farmers who protected environments used by breeding corncrakes. It has also meant much reporting and monitoring by the RSPB, hence the 'Wanted' poster I'd seen at Rapness that morning. It's crucial work and might just save the bird from becoming extinct in Scotland, as has already happened in England and Wales. In Amy Liptrot's superb book *The Outrun*, she describes a summer in her native Orkney spent looking for corncrakes on behalf of the RSPB. She searched for corncrakes each night and earned herself the name 'the corncrake wife'. I couldn't compete with that as I stood stock-still and peered into the long grass around the edges of the field.

Corncrakes are funny-looking birds. Think of a small, scrawny chicken or a young pheasant pullet and you won't be far off. They have long necks and are mainly mottled light-brown in colour. They have a peculiar, scooting run like the cartoon Road Runner and are hard to spot at the best of times, being extremely well camouflaged and often nocturnal.

It was the call that stopped me in my tracks though. It was a repetitive, rasping call, like the noise from moving the wheel on a computer mouse too frantically. The Latin name of the corncrake is *Crex crex*, which must be onomatopoeic. I heard it only a few times, and then it was gone. I never even caught sight of the hidden little bird. It made me smile though, and I couldn't wait to get back into phone signal range and call it in. Corncrakes were alive and well on Westray.

Back at the Bay of Tafts, the slope was still crowded with puffins. One puffin in particular was using a flat stone as a landing strip, and there were others up on the cliffs too. It was pretty playful up there, with puffins pushing each other from ledges and jostling for space. I'd seen a tea towel in a shop in Kirkwall that punned that puffins were 'Orkney Rock Stars' and they were certainly living up to that today. A puffin leant from its perch to examine a clump of pretty sea mayweed, its clown-like make-up painting triangles over each inquisitive eye. Its rainbow beak caught the sun and was perfectly framed by the daisy-like flowers with egg yolk centres.

Further around the clifftop, I eased myself to the edge on all fours and looked down onto the puffins. The occasional fulmar flew past, but the puffins were not concerned. Most were re-waterproofing their feathers, performing head-twisting gymnastics as they preened, a task they would repeat throughout the day. A wheatear came to check on what I was doing.

The striated cliffs stepped down to the beautiful aquamarine sea, tailing off to a natural granite slope. Opposite me, the sea stack was still busy, its countless burrows clearly visible and obviously very well used. I was transfixed by this community of puffins. The stack was scored with so many angular incisions, it looked as if it might tip into the sea at any moment, under the weight of all those puffins. This would probably be my last chance to see puffins that year, but what a way to end the season. It was stunning.

EPILOGUE

Puffin Therapy

There is a chance, of course, that this is all too late. That we've made irreversible mistakes which mean that the climate has changed forever, that the seas are too polluted, and that the puffin, like many other species, is on a crash course to extinction.

This is more than likely what happened to Dow's puffin, a distant relative of the Atlantic puffin. The fossilised remains of these small puffins were found on San Miguel and San Nicolas Islands off southern California (Guthrie, Thomas & Kennedy, 2000). Sadly, the remains of eggs and juvenile birds showed that a number of the Dow's puffins had perished in their burrows, even while parents were sitting on eggs, and the species had evidently failed to reproduce. A sad thought, and something that could happen again.

During my travels, I went to places that would once have been humming with puffins but were now struggling. You don't have to travel to southern California to see this. In Dorset, very few puffins now line the delightfully named Dancing Ledge and there is a worrying scarcity of puffins at South Stack in Anglesey. There are other sites across the UK where puffins used to be found but are now entirely missing. Skokholm's trailblazing conservationist RM Lockley wouldn't believe the trouble his beloved puffins are in.

And yet there remain puffin strongholds where colonies are thriving. In some cases, this is thanks to human intervention, often undoing some earlier wrong, such as on Lundy. In other places, puffins have the perfect conditions to breed and are able to do more than just survive, such as on St Kilda, the Shiant Isles and Skomer, where huge wheels of puffins fill the skies every spring. Perhaps there is some hope left, and it is just change we are seeing, not a charge towards extinction.

I've been astounded by the organisations and individuals working their socks off in order to protect, monitor and further understand our seabirds. From the largest societies to the volunteers on the ground, it's clear that there is a ground swell of wanting to make a change, so that we can protect our delicate seabird populations. An example of this can be found on the Calf of Man, near the Isle of Man, where puffins once used to breed but haven't been recorded as doing so since 1985. In 2016, to compliment a rat eradication programme, the Manx Wildlife Trust placed tens of decoy plastic puffins on the Calf, accompanied by a sound system with puffin calls, in an effort to entice real puffins back to breed. Although it may seem comical, I really hope they find success. No one can say they haven't tried.

There's perhaps just one more story that I ought to add here. The story of Mundi Lundi. Mundi Lundi was scooped up from the streets of Reykjavík after being hit by a car. He was taken in by kind-hearted Ásrún Magnúsdóttir, who quickly realised that the little puffin was blinded and injured through his ordeal. He had sustained a broken beak and a fractured skull. Seeking support from a vet, and showing determination that others may have not, Ásrún was unwilling to let Mundi die, at least not without a fight. She even converted her outdoor hot tub into a (cold) pool for Mundi, and hand-fed him fresh shrimp. Mundi made a partial recovery, but never fully regained his sight. He became a minor celebrity in Iceland, with his own social media and even a book about him. I had the pleasure of meeting Ásrún and Mundi in their home on the outskirts of Reykjavík, where Ásrún told me the story over hot coffee and Mundi Lundi sat contentedly on my lap. He was not in any pain or distress, and I was so impressed with Ásrún's diligent care of this unfortunate little bird. I think the story sums up for me the kindness of humans towards puffins, in stark contrast to the danger we've put them in as a species.

Puffins are themselves a great source of hope. The mere sight of a puffin can lift the most frazzled of spirits. The boat company that took me out to the Treshnish Isles may well have copyrighted the term 'puffin therapy', so I'll need to be careful, but it's clear to me that puffins can do a great deal for our mental health, maybe more so than most other birds. There is genuine excitement at seeing a puffin, as I witnessed countless times – in the chap I met at Fowlsheugh, in Dave the retired mental-health nurse at Amble, in the kids I shared a boat with on the Isles of Scilly. This excitement is reflected in our culture, of course, via Cath Kidston puffin prints, through numerous books, in children's cartoons. But there's something more to it for me. I've never felt more content than when I've been standing in one of the most stunning locations in the UK, the sea churning beneath me, the wind threatening to rip my clothing, and I've seen a resplendent puffin emerge from a sea-thrift-covered burrow a few metres away, completely unconcerned at my presence. It's a magical thing. Just for a few minutes, the stresses and strains of life are no longer there. This is genuine mindfulness; a rare chance to escape the anxieties of the 24/7 global newsfeed and our can-I-get-off world. It's not simply the dopamine hit from crossing locations off a list (puffin-bagging sounds all sorts of wrong, doesn't it?), but rather the well-known serotonin-inducing benefits of spending time up close and personal with nature.

Author Richard Louv conceived the concept of 'nature deficit disorder', that the lack of nature in our day-to-day lives has a negative impact on everything from body weight to eyesight, and especially our mental health. Doctors are issuing 'nature prescriptions' in Shetland in a scheme which is being rolled out across Scotland. Joe Harkness wrote an entire book on the benefits of Bird Therapy. If my own remedy is getting out into spectacular coastal scenery to see a wonderfully enigmatic little seabird, then so be it. I'm more than OK with that. It's a benefit that I hadn't foreseen at the start of this journey – that puffins could help me.

And then there's the unknown. There's still a huge amount we don't know about puffins. We don't know what we don't know, of course, but we're still only just finding out that they use tools and that their beaks glow. We don't really know why puffins travel so far during the winter. We don't understand how or why they come back to the same burrow with the same mate, year after year. Or why they turn up at sex clinics in Hampshire. If there's so much we don't know about them, maybe they'll continue to surprise us. I really hope so. A world without puffins is too heartbreakingly sad to consider. We need to look after our puffins. Every last puffin.

Further Reading

Barkham, Patrick, *Islander* (Granta Books, 2017)

Cleeves, Tim, *RSPB Handbook of British Birds* (Bloomsbury, 2018)

Colgan, Jenny, *Polly and the Puffin* (Little, Brown 2015)

Cowen, Rob, *Common Ground* (Windmill, 2016)

Couzens, Dominic, *The Secret Life of Puffins* (Helm, 2013)

Darlington, Miriam, *Otter Country* (Granta, 2012)

Dunn, Euan, *RSPB Spotlight: Puffins* (Bloomsbury Natural History, 2014)

Fuller, Errol, *The Great Auk: The Extinction of the Original Penguin* (Bunker Hill Publishing, 2003)

Gannon, Angela, *St Kilda: The Last & Outmost Isle* (Historic Environment, 2016)

Harkness, Joe, *Bird Therapy* (Unbound, 2020)

Harvey, Paul, *Shetland Summer Birds* (Shetland Times, 2018)

King, Simon, *Shetland Diaries* (Hodder & Stoughton, 2011)

Kress, Stephen & Jackson, Derrick, *Project Puffin* (Yale, 2015)

Liptrot, Amy, *The Outrun* (Cannongate, 2015)

Lockley, RM & Fisher, J, *Seabirds* (Bloomsbury, 1989)

Lockley, RM, *Puffins* (JM Dent & Sons, 1953)

MacLean, Charles, *Island on the Edge of the World* (Canongate, 2019)

Maxwell, Gavin, *Ring of Bright Water* (Little Toller, 2009)

Murray, Donald S., *The Guga Hunters* (Birlinn, 2015)

Nicolson, Adam, *Sea Room* (HarperCollins, 2004)

Nicolson, Adam, *The Seabird's Cry* (William Collins, 2018)

Perry, Richard, *Lundy: Isle of Puffins* (Lindsay Drummond, 1946)

Soltis, Sue, *Nothing Like a Puffin* (Walker, 2013)

Taylor, Colin, *Life of a Scilly Sergeant* (Arrow, 2017)

Thomson, Amanda, *A Scots dictionary of Nature* (Saraband, 2018)

Williamson, Henry, *Tarka the Otter* (Penguin Classics, 2011)

Wills, Dixe, *Tiny Islands* (AA Publishing, 2013)

Kickstarter Campaign

In the midst of a global pandemic, in May 2020, I decided to launch *Every Last Puffin* on the crowdfunding website Kickstarter. It was not an easy decision and, although I'd used Kickstarter successfully before, I was sure that everyone would be too concerned about Covid-19 to consider backing a book about puffins. I was wrong. The support was overwhelming. Something about puffins, nature or normality intrigued 101 generous individuals to kindly back my project. I am so grateful. Without them this book simply would not exist.

Huge thanks then to my Kickstarter backers: Ali Sheppard, Ally, Ally Lemon, Amanda Graham, Andy Midwinter, Anna, Anna Adams, Arioch Morningstar, Bec Herdson, Blair, C. M. Wallace, Caroline Smith, Carrie, Catherine, Claus Sterneck, Colin Anderton, Creative Fund,

Cymbaline, Daniel, David Roberts, David Swinnerton, Deborah Willott, Dorothy Field, Einar, Ellen Garner Crawford, Faith, Felix and Anna, Gail, Gemma Whitelaw, Grace Hobley, Graham and Jen Maycock, Greg Field, Helen, Helen Danks, Helen Wain, Hollie, Huw, the Iceland Guy, Ingrid Tozer, Jack and Oliver Hart, Jane, James Littlewood, Jayne, Jenny G, Jill Raine, Jill Freeman, the Jolly-Hannahs, Jonas MacArthur, Julie Hunt, Karen Morecroft, Karen Shaw, Karen Shephard, Katrina Gilman, Keri Sleath, Kevin, Kirsty, Laura, Laura Kramer, Lauren Moore, Leah White, Lee Page, Leo and Cole Smith, Lisa, Liz, Liz and Ian Cumberland, Mal, Mark Pinto, Martin, Martin Stephenson, Matt, Meinou, Michael, Michelle Clarke, Nat, Neil, the Nevilles, Nigel and Jane, Óli Fly, Paul, Pauline Wright, Porcha, Rich Bailie, Richard Eriksson, Roger, Samantha Williams, Sara Perry, Selene F, Sharon Ross, the Singh Family, Stuart Whiting, Susie, Tamsin Page, Tess, Tina, Tom Bright, Tracey, Tracy, Trekker_101, Vessela, and Vicki Hsu.

Acknowledgements

Thank you to the following organisations and individuals for their help, inspiration and support: RSPB, Sæheimar Aquarium, RSPB Fowlsheugh, RSPB Bempton Cliffs (especially Kevin and Angela), RSPB South Stack, RSPCA Stapeley Grange (especially Sara Shopland), John Ratcliffe, Natural Resources Wales, The Wildlife Trust of South and West Wales, RM Lockley, Amble Puffin Festival, RSPB Coquet Island, Dave and Dickie on the Steadfast, National Trust Farne Islands, Joe Badcock and the Guiding Star, Isles Of Scilly Seabird Recovery Project, Isles of Scilly Wildlife Trust, the crew of Turus Mara, Blair Villa B&B in Oban, National Trust for Scotland, the mermaid of Staffa, Scottish Seabird Centre (especially James), Isle of May National Nature Reserve, Scottish Natural Heritage, Kate and Caithness Wildlife Tours, Erland Cooper, Sumburgh

Hotel, Sumburgh Head, RSPB Sumburgh Head, The Mousa Boat, Final Checkout, Hermaness National Nature Reserve, Unst Boat Haven, Unst Heritage Centre, Scottish Wildlife Trust, Handa Skua Project, RSPB Rathlin Island, The Landmark Trust, Adam Nicolson, Joe Engebret, Shiant Isles Recovery Project, Bardsey Island Trust, Colin Evans and Bardsey Boat Trips, Ásrún Magnúsdóttir, Andy and Nicola with Go to St Kilda, Dr Annette Fayet, Amy Liptrot, RSPB Orkney, Manx Wildlife Trust, Nick Hill for his map design, Lucy Ridout, Helen Hart, Catherine Blom-Smith and the patient SilverWood team, and Hollie Childe for her wonderful puffin portrait.

Praise for *Iceland, Defrosted*

'A deeply personal tale with a warm-hearted focus'
NATIONAL GEOGRAPHIC TRAVELLER

'Writes with wit and humour. Wonderful descriptions of Iceland as a country and as his home from home'
READER RECOMMENDATION, THE *GUARDIAN*

'A quirky, easy-reading insight into the country'
WANDERLUST

'Get rid of your guidebook immediately and instead bring along Hancox's comical and honest suggestions'
ICELAND REVIEW

Ingram Content Group UK Ltd.
Milton Keynes UK
UKHW010849100323
418370UK00004B/497